Priyavardhini Panneerselvam
John Baliah

Oral microbiome in Oral squamous cell carcinoma

AF757015

Priyavardhini Panneerselvam
John Baliah

Oral microbiome in Oral squamous cell carcinoma

LAP LAMBERT Academic Publishing

Imprint

Cover image: www.ingimage.com

Publisher:
LAP LAMBERT Academic Publishing
is a trademark of
Dodo Books Indian Ocean Ltd. and OmniScriptum S.R.L publishing group

120 High Road, East Finchley, London, N2 9ED, United Kingdom
Str. Armeneasca 28/1, office 1, Chisinau MD-2012, Republic of Moldova, Europe
Managing Directors: Ieva Konstantinova, Victoria Ursu
info@omniscriptum.com

Printed at: see last page
ISBN: 978-620-2-43455-3

ORAL MICROBIOME IN ORAL SQUAMOUS CELL CARCINOMA

INTRODUCTION

According to the American Association for Cancer Research (2023), Cancer is a group of diseases characterized by the uncontrolled proliferation of cells. Ignoring the body's signal to stop, malignant cells multiply to form tumors in organs and tissues or, in the case of blood cancers, crowd out normal cells in the bloodstream and bone marrow. According to American Cancer Society (2023), cancer is a group of diseases in which cells in the body change and grow out of control. Most types of cancer cells form a lump, or mass called a tumor. An updated definition of cancer should include the force of natural selection acting on the initiation and progression of the cancer cell populations contributing to the definition "cancer is a disease of uncontrolled proliferation by transformed cells subject to evolution by natural selection" [(1)]. A subset of cancers that affect the mouth, nose, throat, larynx, sinuses, and salivary glands are known as head and neck cancers. It is the seventh most common cancer worldwide in 2018 [(2)].

Oral squamous cell carcinoma (OSCC) develops from the mucosal epithelium in the oral cavity, pharynx, and larynx and is the most common malignancy in the head and neck [(3)]. It is the most prevalent and most commonly studied oral cancer. Its development is influenced by host-related and lifestyle factors represented by smoking, alcohol abuse, tobacco chewing, and other virus infections namely *Human papilloma virus* (HPV) [(4)]. The squamous epithelium of the oral cavity is continuously exposed to a variety of microbial challenges, at both cellular and molecular levels [(5)].

The presence of OSCC is linked to a distinct microenvironment that includes saliva, site-specific differences in epithelial turnover rate, microbial flora, and varied tissue compositions. It is also influenced by proximity to the external environment. The etiology of several oral illnesses, including OSCC, may be modulated by such a microenvironment. Thus, it would be intriguing to think about how this particular microenvironment affects OSCC disease development given that the microenvironment has the power to influence genetic expression and cancerous activity [(6)].

The oral microbiome is a complex and diverse ecosystem that plays a key role in oral health. It consists of more than 700 species of bacteria, as well as fungi, viruses, and protozoa, which exist in a delicate balance. Disturbances in this balance, often caused by changes in lifestyle, diet, or hygiene, can lead to oral diseases such as periodontitis and dental caries. Recent studies have also suggested a possible link between the oral microbiome and oral squamous cell carcinoma (OSCC), the most common type of oral cancer. Moreover, it is believed that persistent inflammation accounts for one out of every four cases of cancer. An essential task is to determine how this can be related to the aetiopathogenesis of oral squamous cell carcinoma. There is currently strong evidence linking oral squamous cell carcinoma in particular to head and neck cancer, as well as chronic periodontal disease in general. There is a four-fold greater chance of developing head and neck squamous cell carcinoma for every millimeter of lost alveolar bone. Moreover, it is believed that those who have periodontal disease are more susceptible to malignancies with poor differentiation. As a microbial illness, periodontal disease raises the issue, "What is the role of microorganisms in the etiology?" (7).

Based on the literature search drinking alcohol, and cigarette smoking are linked to changes in the gastrointestinal system, and the oral microbiome. It is possible that the microbiome aids in converting an environmental exposure into a carcinogenic impact. Since there aren't many effective systemic treatments for Head and neck squamous cell carcinoma (HNSCC) and local treatment toxicities are frequently high because of the important structures involved, figuring out the microbiological process causing the illness might provide new information on targeted treatments and primary prevention (8).

Many studies have focused on the human microbiome in an attempt to understand the pathophysiology of human illnesses, such as cancer. When compared to other malignancies, oral cancer mortality is high because it is frequently diagnosed in its later stages. Over 40,000 new cases of cancer are diagnosed annually in the USA, where its frequency has climbed during the previous ten years. Furthermore, the pathophysiology of oral cancer is probably complex and not entirely understood (9).

Furthermore, it is unclear whether changes in the oral microbiome are a cause or a consequence of OSCC. Longitudinal studies are needed to clarify the temporal relationship between the oral microbiome and OSCC. In addition, the oral microbiome is a promising area of research for OSCC. Understanding the complex interaction between the oral microbiome and OSCC may lead to new strategies for prevention, diagnosis, and therapy.

ORAL MICROBIOTA – BASICS AND DIVERSITY

The term microbiota was first defined by Lederberg and McCray who emphasized the importance of microorganisms inhabiting the human body in health and disease. This term refers to the entire habitat, including the microorganisms (bacteria, archaea, lower and higher eukaryotes, and viruses), their genomes (i.e., genes), and the surrounding environmental conditions. This definition is based on that of "biome," the biotic and abiotic factors of given environments. It is argued that this is the definition of the word "microbiota" which refers to all microorganisms, including bacteria, viruses, and fungi, that are present in an environment. The collective genome of microorganisms that live in a certain environment, particularly the human body (e.g., skin, gastrointestinal tract), is referred to as the "microbiome" (10).

It has been shown that microbes play a part in regulating and preserving the biological processes of their human hosts. The chance of contracting an illness, including OSCC, may rise when the healthy interaction between microorganisms and the human host is disrupted (dysbiosis), either by infections from pathogenic microorganisms or by external factors such as drinking alcohol or smoking (11, 12).

The oral microbiome is a complex and diverse ecosystem that plays a key role in oral health. There are almost 700 different bacterial species known to exist in the oral cavity, making it a dynamic and diverse ecosystem (13, 14). In contrast to its bacterial equivalent, viral aggregation in the human oral cavity has not been reported very often (15). The connection between carcinogenesis and human microbiota has been shown in several research. Numerous studies have examined the connection between gut microbiota and carcinogenesis in the digestive tract; for instance, *Helicobacter pylori* has been linked to a higher risk of gastric cancer formation (16). *Fusobacterium* and *Salmonella typhi* have been linked to colon and gallbladder cancer, respectively (17, 18, 19). The generation of carcinogenic metabolites, direct antiapoptotic effects, and chronic inflammatory processes may all have a role in the link between microbiota and OSCC onset (20, 21, 22).

Due to the consistent regulation of temperature and humidity, the mouth offers a perfect condition for the growth of numerous bacteria. Hence, a variety of bacteria can be found in the mouth. As a result, the human oral environment creates a complicated system that impacts both health and oral cleanliness (23).

Considering these matters, research has been conducted on the connection between oral microbiota and periodontal disease. Current studies have verified the significant impact of the oral microbiota on the development of periodontal disease. Periodontal disease, which includes gingivitis and periodontitis, is a long-lasting inflammatory condition brought on by an imbalance in the oral microbiota [(24)].

It has been noted that IL-6, an inflammatory cytokine present in saliva, may serve as an indicator of inflammation in the gums and mouth, in terms of oral hygiene and immune response. In a state of good health, there is a diverse combination of bacterial species in the oral microbiota living in harmony [(25)]. Nevertheless, if this equilibrium is disturbed, it can result in the overgrowth of detrimental bacteria like *Porphyromonas gingivalis, Tannerella forsythia, and Treponema denticola*, which can trigger the development of periodontal disease [(26)].

This bacterium infiltrates the gum tissue, triggering inflammation, and possibly resulting in loss of teeth. Tooth decay (dental caries) is also connected to the oral microbiome. *Streptococcus mutans* (*S. mutans*) is a major cause of dental cavities. This bacterium breaks down sugars from food to create lactic acid, which can weaken tooth enamel and cause cavities. Gomez and Nelson's review highlighted the development of the early oral microbiota, the intricate makeup of caries-related microbial communities such as *S. mutans*, and the environmental factors that can disrupt microbial balance and potentially cause disease [(27)].

Saliva plays a role in the oral environment and contributes to the formation of periodontal disease and dental caries. If the quantity and makeup of saliva are not suitable, the equilibrium of the oral microbiota will be disturbed, leading to a higher likelihood of oral diseases like dental caries and gingival diseases. When saliva production decreases, the mouth's ability to clean itself is compromised, making food particles more prone to bacterial decay. As a result, acidic compounds are created, leading to a higher likelihood of dental caries. Furthermore, a disrupted balance in the makeup of saliva can result in a higher proportion of harmful bacteria linked to gingival and periodontal diseases.

Saliva, produced by the salivary glands, supplies the necessary "wetness" for bacteria to thrive. Nevertheless, the saliva that is released in the mouth contains antimicrobial substances. For instance, agglutinins attach to oral bacteria like *S. mutans* and enhance phagocytosis by inducing bacterial clumping. Lysozymes decompose cell walls, leading to cell lysis. Lactoferrin prevents biofilm development by promoting twitching through iron chelation, while peroxidase inhibits bacterial glycolysis, displaying antimicrobial properties. Furthermore, saliva includes secretory IgA that hinders bacterial adherence to the mucosa, along with the secreted mucins MUC5B and MUC7 that clump bacteria together and stop them from sticking to the oral mucosa. Saliva with these elements helps balance the environment of the oral mucosa and a large quantity of oral bacteria, which includes *S. mutans*. Various factors such as stress, getting older, having Sjögren's syndrome, and undergoing radiation therapy for head and neck cancer can all interfere with the normal functioning of the salivary glands. Decreased saliva production from salivary gland hypofunction leads to significant alterations in the oral microbiota. Changes in the oral microbiota can lead to problems in oral health like tooth decay, gum disease, and mouth inflammation, as well as be linked to different systemic diseases.

Reports have indicated links between oral bacteria and disease states like diabetes, the negative impact of *Fusobacterium nucleatum* on colorectal cancer patients, the presence of this bacteria in colorectal cancer tissue from the oral cavity, the connection between periodontal disease and cardiovascular diseases, the reduction of certain bacteria in the oral cavity of rheumatoid arthritis patients, and cognitive decline. While the specific bacteria responsible have not been pinpointed, scientists believe the oral microbiota plays a role in OSCC.

Oral bacteria have been found in different body tissues, indicating a potential link between oral bacteria and the development of diseases. Nevertheless, there is not enough data available on the alterations in the oral microbiota linked to these illnesses. In the upcoming years, the collection of data will establish a database showing the evolution of oral microbiota in individuals with different illnesses and those who are in good health. The examination of oral bacteria can help with the non-invasive identification and early detection of different diseases. The majority of attention in microbial dysbiosis has been directed towards the bacteria, often known as the "microbiome" element. The term "microbiome" has frequently been used interchangeably with the entire microbiota, leading to confusion at times for the reader. While fungi make up a small part of the larger microbiota, they are likely to have a significant impact on dysbiosis, leading to tissue damage in the oral mucosa [28].

Hence, it is crucial to understand the fungal part of the microbiota, also known as the "mycobiome" in relation to health, to determine if mycobiome dysbiosis may be linked to mucosal pathologies. Certain fungi are incorporated into the human microbiota and are present in different human environments. Yet, they make up a minuscule fraction of the total microbial population in comparison to the bacterial portion of the microbiome. In contrast to the bacteriome, the oral mycobiome makes up less than 0.06% and 0.0001% of the microbial population in saliva and dental plaque, respectively [(29)].

Currently, there is insufficient data to fully comprehend the function of oral microbiota and its consequences for both health and illness. The biological community living in the mouth cavity is dynamic and constantly changing. Consequently, monitoring and screening individuals with OSCC may benefit from the ability to identify changes in the microbiota's composition [(30)].

To look at potential relationships between certain patterns of salivary microbiota and OSCC, saliva provides a non-invasive, and accessible tool [(31)].

ORAL MICROBIOTA AND ORAL HEALTH

The microorganisms that typically reside in our mouth are known as the Human oral microbiome, oral microflora, or oral microbiota [32]. The human microbiome is divided into both a core microbiome and a variable microbiome. The core microbiome, present in every person, includes bacteria, fungi, and viruses that are crucial for the typical health and operation of the mouth. On the other hand, the microbiome varies from person to person, depending on their lifestyle and physiological differences is called variable microbiome [33].

The composition of the microbiome changes depending on personal factors and the environment, potentially leading to health risks if the delicate equilibrium is disturbed [34, 35]. The core microbiome stays consistent, including species essential for the optimal operation of the human body. On the other hand, the variable microbiome is constantly changing in response to environmental factors and lifestyle choices [36].

The mouth, which acts as the entrance to the digestive system, is covered with mucous membranes and includes two different types of areas: the hard surfaces of the teeth and the soft tissues of the oral mucosa. The first part of our body that comes into contact with food or drinks is our mouth. Consistent intake of food and drinks may alter the oral environment, leading to lower saliva production, decreased pH levels, periodontal disease, enamel erosion, and cavities [37].

A decrease in saliva production can lower the mouth's defense mechanism, making it easier for bacteria to thrive. A decrease in pH levels can lead to increased acidity in the mouth, which may erode tooth enamel and potentially result in cavities. Interference in the oral environment can lead to periodontal disease, which is then followed by inflammation. Bacteria linked to periodontal disease, like *Porphyromonas gingivalis*, exhibit higher resilience to stress caused by alcohol when compared to other bacteria in the mouth. The mouth is a necessary and lively part of the human body that contains over 774 different types of bacteria. Nevertheless, just 14% are being raised without a specified name, while 32% have not been grown at all, and 54% are both identified and being cultivated. Oral bacteria can lead to various systemic diseases such as endocarditis, chronic respiratory disease, bone infection in children, preterm birth, and coronary artery disease [38].

Bacteria are the primary element of the oral microbiota. The group of oral bacteria consists of six primary phyla: Bacteroidetes, Firmicutes, Fusobacteria, Spirochaetes, Proteobacteria, and Actinobacteria. The highest density of microbes in the human body is found in saliva, dental plaque, and the oral mucosa. This happens due to saliva containing many nutrients that nourish bacteria. Dental plaque is sticky and accumulates on teeth, creating an environment for bacteria to thrive. About 774 bacteria and roughly 100 types of fungi are believed to live in the mouth. Numerous research studies explore the potential connection between the oral microbiome and systemic disease (39). Therefore, it is essential not to overlook the significance of oral health, as it can impact overall well-being (40). The human body's main entrance is through our mouth (41, 42).

This is due to the fact that the mouth harbors diverse microbial populations that have the potential to migrate to different areas of the body, leading to illness. Furthermore, the mouth has mucous membranes that can be invaded by bacteria and viruses, enabling them to enter the body and lead to infection. The mouth, also known as the oral cavity, harbors a wide variety of microorganisms that can lead to infection if they enter the bloodstream. These tiny organisms create toxins and enzymes that have the potential to harm tissues and organs in the body, resulting in various conditions like cavities, gum disease, and systemic illnesses (43,44).

The oral microorganisms interacting with the host factors in the oral cavity are known as the oral ecosystem (45). The term 'microbial homeostasis' was originally proposed by microbial ecologist Martin Alexander [1971]. Oral microbial homeostasis is the ability to maintain a stable microbial community through dynamic microbe interactions. Oral allostasis refers to the adaptation of the oral ecosystem to changing conditions. Puberty, which usually lasts between 2 and 4 years, is an example of Allostasis.

Maintaining oral health depends on the commensal oral microbiota. Commensals outcompete pathogenic species for colonization substrate, limiting the opportunity for integration by foreign pathogens. This is one way that commensals support oral health. Oral commensals that exhibit direct antagonistic behaviour towards oral pathogens include *Actinomyces naeslundii, Haemophilus parainfluenzae, Streptococcus salivarius, Streptococcus mitis, Streptococcus sanguinis, and Streptococcus cristatus* have all been demonstrated to reduce *Porphyromonas gingivalis* capacity to attach to substrate.

Furthermore, clinical isolates from people with good oral health included *Streptococcus, Actinomyces, and Bifidobacterium* strains, all of which were found to suppress *Porphyromonas gingivalis* growth. *Lactococcus lactis*, an oral microbiota commensal, generates nisin, a bacteriocin that has proven to prevent oral carcinogenesis and increase the lifespan of tumor-bearing mice. Furthermore, nisin was recently shown in vitro to reduce pathogen-mediated oral carcinogenesis, cancer cell migration, and cell invasion of oral squamous cell carcinoma, indicating that commensal species may have an anti-cancer function. In addition to nitrate metabolism, the oral microbiota plays a role in systemic nutrient cycling [(46)].

Nearly 25% of ingested nitrate is transported, via the enterosalivary circuit, to the oral cavity and these oral microbes reduce nitrate to nitrite, which is taken into the bloodstream during digestion and converted to nitric oxide. Nitric oxide is important to cardiovascular health as it exerts vasodilation and antihypertensive effects. Such findings demonstrate the importance of the commensal oral microbiome in maintaining oral health and promoting systemic health. Dysfunction of nitric oxide signaling is associated with pulmonary hypertension, obesity, and cardiovascular disease. Studies have found that the use of antibacterial mouthwash can increase blood pressure as a result of its inhibitory effect on the oral microbiome [(47)].

The interactions between various microbial species in the biofilm determine changes in the microbial biomass in the oral cavity. The oral biofilm develops as a result of interactions between early colonizing bacteria and later colonizers via a variety of pathways, including metabolic exchange, coagulation, small-molecule signal-mediated communication, and genetic material exchange. The many bacteria that inhabit the oral cavity are not evenly dispersed across all surfaces; rather, their metabolism in a healthy mouth cavity determines how they multiply in distinct ecological niches.

The ecology of these habitats is influenced by various oral anatomical locations, which result in microenvironmental variances. *Streptococcus, Gemella, Eubacterium, Selenomonas, Veillonella, Actinomyces, Atopobium, Rothia, Neisseria, Eikenella, Campylobacter, Porphyromonas, Prevotella, Granulicatella, Capnocytophaga, Fusobacterium, Leptotrichia, Streptococcus mitis, and Granulicatella adiacens* were the most prevalent species of buccal epithelium. These pathogens are frequently regarded as opportunistic pathogens and were found in bacteremia / septicemia in patients with infectious endocarditis/atheroma.

A variety of species, including *Granulicatella adiacens, Veillonella spp., Streptococcus mitis, Streptococcus australis, Streptococcus parasanguinis, and Streptococcus salivarius*, were found on the tongue dorsum between keratinized papillae. The bacterial profiles were impacted by anatomical variations between the tongue's dorsum and lateral border. The smooth, nonkeratinized surface of the tongue's lateral margin is home to some bacteria, including *Streptococcus mitis, Streptococcus mitis bv2, Streptococcus sp.* clones DP009 and FN051, *Streptococcus australis, Granulicatella adiacens, Gemella haemolysans, and Veillonella* spp. The species of tongue most closely linked to health were *Rothia mucilaginosa (Stomatococcus mucilaginosus), Streptococcus salivarius*, and an unidentified species of cultivable *Eubacterium* (strain FTB41).

Similar to this, the predominant bacterial species on the hard palate were *Neisseria subflava, Granulicatella elegans, G. hemolysans, Streptococcus mitis, and Streptococcus mitis biovar2*. The most frequently discovered species in nearly all healthy patients and settings was *Streptococcus mitis*. On the other hand, endocarditis caused by bacteria has been linked to both *Streptococcus oralis and S. mitis* in individuals who had prosthetic valves. Apart from the unique bacteria found in a healthy oral cavity, there were also microorganisms linked to periodontal disease, including *Treponema denticola, Tannerella forsythia, and Porphyromonas gingivalis* [(48)].

The host and its obligatory or facultative symbionts, whose dynamic interconnection requires emergent interactions and results, can be included in the holistic framework of holobiont, which also includes the interspecies and host-microbial interactions [(49)].

Currently, available multi-omics techniques allow for the investigation of the intricacy of interspecies interactions, polymicrobial habitats, and the processes behind these connections [(50)].

HEAD AND NECK SQUAMOUS CELL CARCINOMA - EPIDEMIOLOGY AND PREVALENCE

In this comprehensive review, we aim to delve into the epidemiology of HNSCC, exploring various aspects such as the incidence and mortality rates associated with this malignancy. By examining the geographical and demographic factors that contribute to HNSCC disparities, we can gain insights into the underlying mechanisms and risk factors involved. Additionally, we will thoroughly examine the most common risk factors associated with HNSCC. These risk factors range from well-established ones such as tobacco, smoking, and alcohol consumption to emerging factors such as viral infections, including human papillomavirus (HPV), and Epstein–Barr virus (EBV) infections. Understanding the multifaceted nature of these risk factors are crucial for devising effective prevention and intervention strategies that can help reduce the incidence of HNSCC. Furthermore, evidence-based prevention strategies for HNSCC encompass various approaches, including lifestyle modifications, early detection methods, and vaccination against high-risk viral infections.

Global prevalence

According to the most recent GLOBOCAN data (2020), HNSCC ranks as the seventh most prevalent cancer worldwide, with around 890,000 new diagnoses (approximately 4.5% of all cancer cases globally) and 450,000 deaths annually (roughly 4.6% of worldwide cancer-related fatalities). Around 380,000 cases of lip and oral cavity cancer, 185,000 cases of larynx cancer, 133,000 cases of nasopharynx cancer, 98,000 cases of oropharynx cancer, 84,000 cases of hypopharynx cancer, and 54,000 cases of salivary gland cancer have been reported [51].

The frequency and death rates of HNSCC differ greatly based on geographic location and demographic factors. On a worldwide scale, HNSCC is more prevalent in males compared to females, with a ratio of about 2:1, and in individuals above the age of 50 [52].

The highest prevalence of HNSCC is found in South and Southeast Asia due to the common practice of chewing the cancer-causing areca nut. Central and Eastern Europe, as well as South America, also have high incidence rates [53]. In India, the highest rates of occurrence are seen in cases of HNSCC, where tobacco (with or without the areca nut) is responsible for as much as 80% [54].

The worldwide prevalence of HNSCC is on the rise in several nations, especially among younger demographics, and is expected to increase by 30% annually by 2030 [54]. Changes in lifestyle, including higher alcohol and tobacco use in developing countries, along with the increasing cases of HPV-related oropharyngeal cancer, are partially responsible for this trend [55].

HPV is projected to become the primary cause of HNSCC globally, surpassing tobacco, and leading to a higher incidence of oropharyngeal HNSCC compared to oral cancer, which is mostly linked to tobacco use. Likewise, in the last ten years, there has been a 23% rise in the number of laryngeal cancer cases [56]. In developed countries, there has been an increase in the number of younger women experiencing higher rates of disease, possibly caused by shifts in sex-specific cultural norms around tobacco and alcohol use, along with a rising burden of HPV [57].

In Japan, there has been an increase in the occurrence of tumors in the oropharyngeal, oral cavity, and salivary glands, particularly among women, while there has been a decrease in the incidence of tumors in the nasopharynx and larynx, which are more linked to the Epstein-Barr virus in East Asian populations rather than HPV [58]. Similarly, there has been a 7.3% increase in oropharyngeal HNSCC among men and a 6.5% increase among women in the UK. Additionally, oral cancer rates have risen by 2.8% for men and 3.0% for women, particularly impacting those with lower socioeconomic status [59, 60].

Indian prevalence

Bagal et al (2022) stated that the incidence of HNSCC in all sites was about 103.7 and 102.4 per 1 lakh population. The occurrence of cancer was found to be increased with increase in age. The age standardized incidence rate (ASIR) for the age group 40-59 years was in the range of 36.3-62.6 and 12.4-20.8 per 1 lakh population in males and females, respectively. The incidence of head and neck cancer in Northern India is about 28.5 and for central India, it is about 28.3 per 1 lakh population. The highest incidence is seen in males 31.7 and for females it is about 6.2-10.1 per 1 lakh population.

Every 1 in 33 males and 1 in 107 females in India are at risk for developing head and neck cancer. The age standardized mortality rate (ASMR) of head and neck cancer in India as per GLOBACAN 2020 for males is 14.2 and for females is 4.1 per 1 lakh population [61].

In India, oral cancers rank third most common cancer and gingivo-buccal complex (alveolar ridge, gingival-buccal sulcus, buccal mucosa) forms the most common sub-site for cancer of the oral cavity in contrast to the tongue and floor of the mouth that is more common in the western world [(62)].

Five-year absolute survival in patients with OSCC according to the Mumbai registry before 2000 were 62.5%,18%, and 1.1%, respectively, for localized, regional, and metastatic disease; the figures according to the Chennai registry were 54.9%, 30%, and 10%, respectively [(63)].

Simi et al (2021) conducted a descriptive study to analyze the burden of OSCC in Kerala. 243 histologically proven cases of OSCC were used as the sample for the study with the age range from 19 to 92 years. Almost 95% of the cases were 40 years of age with the male to female ratio of about 1.4:1. As OSCC has male gender predilection due to tobacco and alcohol habits, in India the incidence of OSCC has increased due to the use of smokeless tobacco by women [(64)].

ORAL MICROBIOME AND ORAL SQUAMOUS CELL CARCINOMA

A comprehensive review by **Charlotte Thomas et al** (2021) on oral microbiota in the diagnosis of systemic diseases stated that there was a physio-pathological mechanism linking oral dysbiosis to systemic diseases as depicted below [(65)].

Oral Dysbiosis including microbial alteration in both saliva and Periodontium

↓

Local inflammation leading to release of TNF-a and Interleukins

↓

Blood stream

↓

Interacts with the bacterial lipopolysaccharides/teichoic acid/Flagellin

↓

Target organs and reacts with the toll like receptors

↓

TNF-a and pro-inflammatory Interleukins production

↓

Systemic Inflammation

↓

Systemic disease

Lu Gao et al (2018) reviewed the importance of oral microbiota in the oral cavity. The healthy oral conditions are primarily determined by the proper equilibrium of the different types of bacteria in the oral microbiome, and when this balance is disrupted, dysbiosis occurs. Dysbiosis arises when certain pathogenic microorganisms multiply excessively due to factors like antimicrobial agents, age, hormonal changes, poor oral hygiene, smoking, or other health conditions. Dysbiosis can result in the development of several oral disorders, such as cavities, gum disease, and inflammation of the gums. Both periodontitis and gingivitis have inflammation as a common feature, however, periodontitis is considered a more severe condition compared to gingivitis. In contrast, gingivitis only affects the soft tissues around the teeth, while periodontitis is a long-term condition that causes damage to the structures supporting the teeth and is known as the primary reason for tooth loss in adults (66).

In a case-control study on oral microbiome using stimulated saliva, the authors observed that bacteria related to gum disease were higher in the mouth of people with oral cancer compared to those with healthy individuals. These bacteria include *Porphyromonas gingivalis, Fusobacterium nucleatum, Pseudomonas aeruginosa, Fusobacterium periodonticum, Aggregatibacter segnis, Campylobacter rectus, Campylobacter showae, Peptostreptococcus stomatis, Peptostreptococcus micros,* and *Catonella Morbi.* On the other hand, the bacteria *Streptococcus, Veillonella, and Rothia* were lower in the cancer tissue. Losing teeth and having a disease can lead to fewer types of bacteria in the mouth. Most of the harmful bacteria that cause gum disease need little or no oxygen to survive, and their numbers go up a lot in the low-oxygen environment of the tumor. Also, the bacteria in the tumor tissue were able to break down sugar and survive in acidic conditions. Some types of bacteria can change the proteins in our bodies into harmful substances, which can cause cell mutations. The type of bacteria in our mouths can also change depending on how we take samples and on the stage of oral cancer (67).

Hubber et al (2015) reviewed the association of human papillomavirus and oral cancer and concluded that several studies had shown that the HPV DNA was usually present in a considerable number of OSCC cases. However, there was a lack of clear molecular evidence that HPV DNA-positive OSCC is caused by HPV since the presence of DNA of the virus does not mean the presence of active human papillomavirus (68).

Adler I et al (2014) elicited the importance of *H. pylori* in the development or progression of cancer, leading to the International Agency for Research on Cancer of the World Health Organization classifying *H. pylori* as a human carcinogen. The presence of *H. pylori* has also been reported in dental plaque and saliva. The authors concluded that *Helicobacter pylori* has a relation to different oral mucosal pathologies like periodontal disease, canker sores, burning tongue, and its association with gastric pathology (69).

Xue Meng et al (2016) conducted a case-control study with OSCC patients and concluded that there is a positive correlation between *H. pylori* infection and OSCC occurrence. In addition, there was a negative correlation in people older than 60 years [(70)].

Belibasakis et al (2024) explained the dysbiosis of bacteriome and mycobiome in oral dysplasia and oral cancer. The possible contribution of fungi and bacteria in OSCC development was depicted below.

Bacteria enter the bloodstream

↓

Translocated from the periodontal pocket and during periodontal microinjury leading to

- Proinflammatory microenvironment
- Alteration of the balance between host cell proliferation and apoptosis
- Tumor Microenvironment created [(71)]

Lamont et al (2020) elucidated the role of *P. gingivalis* in OSCC. It is justifiable to inquire if changes in the extensive and varied makeup of the oral microbiome, or the continuation of oral bacterial infection, might serve as a trigger for OSCC. Frequently, changes in the microbial community can be seen in typical oral illnesses like periodontitis and caries, caused by a disturbance in the balance of multiple microorganisms and an increase in harmful bacteria. Pathogenic bacteria in the mouth, like *Porphyromonas gingivalis*, have been linked to oral precancerous lesions and OSCC [(72)].

Vyhnalova et al (2021) described the etiopathogenesis of OSCC and the correlation might be related to the fact that oral infections are naturally occurring and oral bacteria can impact inflammation. The imbalanced microbiome present in the tumor microenvironment probably plays a role in tumor advancement by maintaining long-term inflammation. Therefore, the upcoming sections will offer information on the microbial communities associated with OSCC discovered in different sample sources such as oral rinses, saliva, plaque, tumor swabs, biopsy sections, and tissue homogenates. Tumor swabs and homogenates offer proof of bacterial presence in or invasion of the tumor site. On the other hand, utilizing immunohistochemistry to stain biopsy sections can show the bacteria's intracellular location [73].

X.Yu et al (2023) performed a systematic review and meta-analysis on dysbiosis of microbiota in OSCC. The bacteria depleted in OSCC include *Streptococcus, Haemophilus, and Neisseria and* were in close association with periodontitis. *Fusobacterium nucleatum* and *Prevotella* bacteria were found to be significantly increased in patients with OSCC [74].

INTERACTION BETWEEN ORAL MICROBIOTA AND HOST CELLS

The mechanisms by which the oral microbiota might contribute to OSCC development include increasing oxidative stress, facilitating local invasion and metastasis, and promoting cell proliferation. The summary of some oral microbiota and their association for OSCC development is mentioned below [75].

- CARCINOGENIC SUBSTANCE PRODUCTION
  - ✓ *Porphyromonas gingivalis*
  - ✓ *Fusobacterium nucleatum*
  - ✓ *Prevotella intermedia*
  - ✓ *Streptococcus species*
- CELLULAR PROLIFERATION AND ANTIAPOPTOTIC ACTIVITY
  - ✓ *Porphyromonas gingivalis*
  - ✓ *Fusobacterium nucleatum*
  - ✓ *Human papilloma virus*
- CELLULAR INVASION
  - ✓ *Bacteroides fragilis*
  - ✓ *Epstein Barr virus*
  - ✓ *Candida albicans*
- INFLAMMATORY AND IMMUNE RESPONSE REGULATION
  - ✓ *Porphyromonas gingivalis*
  - ✓ *Fusobacterium nucleatum*
  - ✓ *Prevotella intermedia*
  - ✓ *Tannerella forsythia*
  - ✓ *Treponema denticola*

***Porphyromonas gingivalis* in Oral Cancer**

The category of *Porphyromonas* consists of bacteria that are Gram-negative, asaccharolytic, obligately anaerobic, non-spore-forming, non-motile, and pleomorphic bacilli (76). A member of the Porphyromonadaceae family is the species *P. gingivalis* found in this genus. This type of bacteria is frequently present in the deep periodontal pockets of humans. It is one of the most important periodontal pathogens present in the oral cavity associated with an increased risk of malignancies such as OSCC, esophageal squamous cell carcinoma, and gingival carcinoma. Sayehmiri et al (2015) observed that this pathogen increases the risk of OSCC by 1.36-fold (77).

The carcinogenic mechanisms of various bacteria are listed in Table 1.

Table 1. Oral bacteria and their carcinogenic mechanism

ORAL BACTERIA	CARCINOGENIC MECHANISM
C. gingivalis	✓ Induction of EMT through downregulation of epithelial markers and upregulation of mesenchymal markers and transcription factors
Lactobacillus	✓ Production of lactic acid and oncometabolites such as hydrogen peroxide
F. nucleatum	✓ Overexpression of metalloproteinases induced by p38 MAPK activation ✓ Induce a robust inflammatory response by activation NF-kB and NLRP3 inflammasome ✓ Induction of EMT via the lncRNA MIR4435-2HG/miR-296-5p/Akt2/SNAI1 signaling pathway
P. stomatis	✓ Production of several types of acids, including lactic, acetic, butyric, isobutyric, isovaleric, and isocaproic acids

P. gingivalis	✓ Inhibition of apoptosis by the production of nucleoside diphosphate kinase, stimulation of Jak1/Stat3 signaling pathway and upregulation of miRNA-203 ✓ Induction of chronic inflammation through IL-6, IL-8, TGF-b1, TNF-a and B7- H1receptor expression ✓ Production of oncometabolites such as oxygen radicals, butyrate, and acetaldehyde
P. intermedia	✓ Induction of chronic inflammation by the production of fimbriae adhesins, LPSs, peptidoglycans, and lipoteichoic acids ✓ Activation of PARs by secretion of proteases ✓ Production of oncometabolites such as hydrogen sulphide and methyl mercaptan
S. anginosus	✓ Induction of DNA Damage through increased synthesis of NO and COX-2
S. gordonii	✓ Induction of EMT through decreasing ZEB2 expression ✓ Production of oncometabolite such as acetaldehyde
S. mitis	✓ Suppression of OSCC cell proliferation by the production of hydrogen peroxide ✓ Production of oncometabolite such as acetaldehyde
T. denticola	✓ Induction tumor invasiveness by dentilisin overexpression

P. gingivalis induces OSCC through various mechanisms. *P. gingivalis* releases nucleoside-diphosphate-kinase (NDK), which enhances tumor formation. NDK blocks the ability of ATP to activate purinergic receptor (P2X7) receptors, leading to decreased IL-1b production in epithelial cells. IL-1b is necessary to prepare IFN G, leading to the development of tumor-specific CD8+ T cells.

Therefore, *P. gingivalis* NDK could assist tumors in evading immune detection. NDK-mediated ATP degradation inhibits apoptosis by suppressing P2X7 receptor activation that requires ATP. Moreover, *P. gingivalis* NDK also phosphorylates heat shock protein 27

(HSP27) to impart an antiapoptotic phenotype to primary gingival epithelial cells. Yilmaz and colleagues discovered that *P. gingivalis* prevents epithelial cell death, which has dual effects on cancer - it can promote carcinogenesis while also protecting malignant cells. The progression can be explained by the activation of the Jak1/Akt/Stat3 signaling pathway, an elevated Bcl2 (anti-apoptosis) to Bax (pro-apoptosis) ratio, decreased cytochrome c release, and the blocking of caspase-9 and executioner caspase-3. It also controls the expression of microRNAs and increases miR-203 expression, leading to decreased levels of SOCS3, resulting in inhibition of epithelial cell death. SOCS3 prevents JAK/STAT3 signaling by attaching to phosphorylated JAK receptors.

It results in enduring intracellular infections within epithelial cells. Chronic inflammation has been associated with cancer due to the production of inflammatory molecules such as IL-6, resulting in DNA hypomethylation and promoter hypermethylation. Metabolites produced by this bacterium could be a potential factor in the development of cancer. Metabolites include oxygen radicals, butyrate, and acetaldehyde. Butyrate, a potent bacterial carcinogen, can induce apoptosis in T- and B-lymphocytes and impair their function. Acetaldehyde, a harmful byproduct of ethanol metabolism, has the potential to harm DNA and encourage abnormal growth of epithelial cells. The toxic metabolites produced by this bacterium could result in the OSCC [(78)].

***Fusobacterium nucleatum* in OSCC progression**

Within the genus *Fusobacterium* there are various species of Gram-negative rods that are obligate anaerobes, do not form spores, and may or may not be motile. *Fusobacteria* are recognized as bacteria that colonize later. The appearance of OSCC has been linked to the Fusobacterium nucleatum bacteria that live harmoniously in the human mouth. The biofilms found on OSCC are filled with anaerobic periodontal pathogens, like *F. nucleatum*. This has prompted the idea that these bacteria could have a role played a role in the advancement of oral cancer. By utilizing various methods such as creating toxins, enzymes, and signaling molecules, this bacterium can adhere to the oral epithelium, provoke inflammation, change the immune response, and promote the development of cancer.

F. nucleatum links its RadD adhesin with the streptococcal adhesin SpaP to interact with *Streptococcus* spp., the initial colonizers of dental surfaces. This interaction also includes Aid1 and CmpA, two more fusobacterial adhesins. *F. nucleatum's* adhesion proteins RadD,

Fap2, and FomA attach to other bacteria like *Porphyromonas gingivalis* in the biofilm. The connection between *F. nucleatum* and *Actinomyces naeslundii* and *Candida albicans* is also shown by RadD. *Fusobacterium* plays a role in the development and advancement of OSCC. *F. nucleatum* infection leads to the activation of collagenase 3, an enzyme that breaks down extracellular matrix and helps in tissue breakdown and cell movement. Infection by *F. nucleatum* causes NF-kB to move into the nucleus, leading to the production of cytokine genes. Both oral epithelial cells and OSCC cells display this mechanism. When F is present, cells undergo a process called epithelial-mesenchymal transition (EMT). The EMT indicators, including E-cadherin, N-cadherin, vimentin, and snail family transcriptional, were impacted by the infection [(79)].

Gemella

Gemella species are cocci that are Gram-variable and can be facultatively anaerobic, and they have the ability to group together in pairs, clusters, or short chains. These species are recognized for occasionally causing systemic illnesses. In some study comparing the oral microbiota of individuals with OSCC, researchers found a close connection between the location of the tumor and two bacterial species, *Gemella haemolysans* and *Gemella morbillorum,* which are part of the oral microbiome in humans. Additional investigation through metagenomics showed that the microbiome underwent changes depending on the depth of invasion (DOI) of OSCC. More precisely, the levels of various bacteria, such as *Gemella haemolysans* and *Gemella morbillorum*, grew as DOI increased, whereas other bacteria saw a decline in abundance. Some research indicates that particular Gemella species may play a role in the development and dissemination of OSCC.

Other bacteria

In addition to the above-mentioned bacteria, there are several other types of oral bacteria that are also carcinogenic. Compared to healthy individuals, samples from oral cancer patients showed a higher diversity of bacterial species. Specifically, OSCC samples taken from individuals with pre-existing periodontitis showed overexposed levels of *Catonella, Dialister, Filifactor* and *Parvimonas* in the samples. Significant differences in microbial composition and profile were observed between matched Para cancerous tissues with OSCC lesions. Bacteria from the genera *Carmanella, Tannerella, Parvimonas* and *Filifactor* were associated with malignant tumors [(80)].

Viruses

The carcinogenic mechanisms of various viruses are listed Table 2 [(79)].

Table 2. Carcinogenic mechanism of viruses

VIRUS	CARCINOGENIC MECHANISM
Epstein Barr virus	Activating different signaling pathways, such as NF-kB, PI3K/AKT, JAK/STAT, MAPK, TGF-b, and Wnt/b-catenin by LMPs, EBNAs and EBV miRNAs
Human papilloma virus	Prolongation cell cycle, activation cellular proliferation and prevention apoptosis by E6 and E7 oncoproteins

Around 20% of malignant neoplasms in humans are linked to viral or bacterial infections. Peyton Rous discovered that viruses have the capability to induce cancer. He showed that a virus called *Rous sarcoma* virus could transmit chicken sarcomas. Now, it is known that *Epstein-Barr* virus, *hepatitis B* virus, *hepatitis C* virus, *human papilloma* virus, *human T-cell lymphotropic* virus, and *Kaposi's associated sarcoma* virus contribute to 10-15% of global cancer cases. HPV *i*s widely recognized as a cancer-causing factor in the development of cervical cancer. Many studies have backed the idea of viruses playing a significant role in the onset of oral cancer, particularly in relation to OSCC and precancerous oral lesions. Various types of HPV are recognized for their strong cancer-causing abilities, as shown in multiple research studies detailing the process by which HPVs lead to and sustain the development of malignancies. The primary proof that HPV plays a role in cancer development is that their genes and gene products can disrupt the cell cycle. HPV contains two main oncoproteins, which are the E6 and E7 proteins. The E6 and E7 proteins use methods to decrease the expression of histocompatibility genes, impacting both innate and adaptive immunity. Many studies show a growing amount of proof of HPV-related oral cavity cancer in men. It has also been shown that HPV is found in healthy oral tissue, indicating that its involvement in oral cancer development is uncertain. A study of sixty confirmed OSCC samples was conducted, where HPV was tested using polymerase chain reaction. Research has indicated that all samples tested negative for HPV. Other researchers employed ultra-high-throughput sequencing of the cancer transcriptome to determine the presence of papilloma virus transcripts in cutaneous squamous cell carcinoma [(81)].

Fungi

The carcinogenic mechanisms of Fungi are listed in Table 3 [(79)].

Table 3. Carcinogenic mechanisms of fungi

FUNGI	CARCINOGENIC MECHANISM
Candida albicans	✓ Damaging the mucosal epithelium ✓ Producing carcinogens such as nitrosamines ✓ Triggering chronic inflammation ✓ Inducing Th17 immune responses

Monteiro et al (2024) observed that candidiasis triggers type 3 responses initiated by activated IL-17-secreting cells. These responses have been linked to tumor-promoting infiltrates of inflammatory cells and pose an antagonistic action to interferon-gamma (IFN-γ), a recognized anti-cancer defense mechanism. Conversely, candidalysin, a toxic epithelium-damaging substance produced by *Candida*, has been postulated to enhance tumor-promoting immunity. Despite this, the direct link between candidalysin and oral carcinogenesis remains unestablished [(82)].

This recently identified *Candida* virulence factor has been demonstrated to activate additional oncogenic pathways within epithelial elements, involving the epidermal growth factor receptor/mitogen-activated protein kinase axis. In a study on pancreatic cancer, the mannose-binding lectin, which binds specifically to the glycans on the fungal wall, was activated mainly by the *Malassezia* genus, showing potential for tumor formation. On the other hand, the presence of *Malassezia sp.* in OSCC is associated with a favorable prognosis. While the oral mycobiome has demonstrated carcinogenic implications, substantial proof is currently lacking to establish whether these findings can be replicated for oral epithelial dysplasia. An alternative explanation posits that while these occurrences do exist in oral cancer growth, the associated changes may be insufficient to solely induce dysplastic alterations in oral epithelial tissue, serving more as accelerators rather that primary causes to mimic experimental oral carcinogenesis development, 4-nitrioquinolone-1-oxide (4NQO) has been used. This compound induces mutagenic changes in DNA similar to those triggered by tobacco [(83)].

Live *C. albicans* has been found to enhance the migration potential, induce matrix metalloproteinase, and release associated metabolites via epithelial-mesenchymal shift, along with the upregulation of metastasis-associated genes. The same study explored the capacity of *Candida* to facilitate sequential cancer in 4NQO mice. These pre-neoplastic growths are supported by dysbiosis in the oral mycobiota, allowing them to progress into cancer. Antifungal therapy alone proves inadequate in treating the affected area and addressing this complex process; hence, the lesion should be treated as an OPMD.

In a recent metaproteomic analysis, a total of 196 fungal proteins demonstrated significant changes in abundance across different forms of OSCC. Most of the detected fungi had not been thoroughly investigated, and researchers were unaware of their toxicity or lethal associations in the oral cavity of humans. However, other fungi, including *Verruconis gallopava, Dimargaris cristalligena*, and *Syncephalastrum racemosum*, were found to be more frequently present in OSCC lesions and linked to opportunistic infections. Additionally, other well-known fungi species such as *Paracoccidioides brasiliensis*, *Malassezia sympodialis*, and *Lichteimia corymbifera* were detected. In an investigation comparing tissue from tumors with non-tumor controls, *L. corymbifera* showed an association with OSCC of the tongue. *P. brasiliensis*, a fungal yeast causing *paracoccidioidomycosis* leading to oral manifestations, has been associated with OSCC. Coexistence of oral *paracoccidioidomycosis* and OSCC has been reported, suggesting a potential role of fungi in cancer development, possibly due to continuous stimulation of epithelial cells increasing susceptibility to malignant transformation [(84)].

Di Cosola et al (2021) proposed several molecular mechanisms by which *C. albicans* can cause both the oral premalignant and malignant condition in oral cavity which include

- Production of nitrosamines
- Overexpression of factors like p53, Ki-67, and COX-2
- Overproduction of acid aspartyl proteinase
- Upregulation in proinflammatory cytokine like IL-1, IL-6, IL-8
- Reduction of β defensins
- Cytolytic toxin candidalysin [(85)]

Weiwei Heng (2022) et al analyzed the oral bacteriome and mycobiome along the health-premalignancy-carcinoma sequence. Some species, including *Prevotella intermedia, Porphyromonas endodontalis*, *Acremonium exuviarum*, and *Aspergillus fumigatus*, were increased, whereas others, such as Streptococcus salivarius, Scapharca broughtonii, Mortierella echinula, and Morchella septimelata, were depleted in OSCC (86).

CLINICAL EVIDENCE AND ASSOCIATIONS OF ORAL MICROBIOME WIH ORAL CANCER

Chattopadhyay et al (2019) reviewed the diagnosis and prognosis of oral cancer and explained that early diagnosis is essential. Identification and application of saliva-based genomics, proteomics, metagenomics, and metabolomics derived biomarkers may overcome painful invasive procedures. Accumulation of both genetic and epigenetic changes, which lead to a gain of function in certain oncogenes and a loss of function in some tumor suppressor genes of oral carcinogenesis, are well documented. "Omics" biomarkers in saliva, such as metabolites, proteins, transcribed genes, miRNAs, genome alterations, and epigenomic changes, as well as the microbiome have been proposed for use in the diagnosis of oral cancer.

The various dysregulated salivary mRNAs and their association with OSCC are listed in Table 4.

Table 4. Dysregulated salivary mRNAs and their association with OSCC

Gene Symbol	Gene Name	Functional Importance in Cancer
DUSP1	Dual specificity protein phosphatase 1	Responsible for Chemoresistance and radioresistant properties in solid tumors
H3F3A	H3 Histone Family 3A	Proliferation of cells
IL-1β	Interleukin 1 Beta	Inflammation, cell proliferation, differentiation, and apoptosis
IL8	Interleukin 8	Tumor angiogenesis, cell adhesion, immunity, and cell cycle arrest
SAT1	Spermidine/Spermine N1-Acetyltransferase 1	Cell Proliferation and angiogenesis
OAZ1	Ornithine decarboxylase antizyme 1	Cell growth and proliferation by controlling intracellular levels of polyamine.

S100 P	S100 calcium binding protein P	Cell cycle regulation and differentiation
MMP-1	Matrix Metallopeptidase 1	Associated with the progression of dysplasia to cancer through enhancement of susceptibility to angiogenesis
MMP-9	Matrix Metallopeptidase 9	Associated with the progression of dysplasia to cancer through enhancement of susceptibility to angiogenesis.

The commonly deregulated salivary miRNAs and their association with OSCC are listed in Table 5.

Table 5. Deregulated salivary miRNAs and OSCC association were listed below

miRNA	Functional importance in cancer
miR-125a	Inhibits cell proliferation
miR-200a	Tumor suppression and early metastasis
miR-31	Enhance cell proliferation, promotes metastasis, monitoring or detecting residual or recurrent tumor
miR-136	Tumor-suppression and promotes apoptosis.
miR-27b	Inhibition of cell proliferation through inhibiting FZD7 and FZD7-mediated Wnt signaling pathway
miR-139-5p	Cell proliferation, apoptosis, lymph node involvement and metastasis, metastatic potential of the tumors, a poor prog
miR-197	Cell proliferation
miR-let-7	Metastasis
miR-99a/b	Tumor suppression
miR-126	Tumor suppression

miR-145	Decreased proliferation, or promoted apoptosis by targeting K-RAS, c- Myc and DFF45
miR- 21	Tumorigenesis and invasiveness through the Wnt/β-catenin pathway, by targeting DKK2
miR-184	Inducing proliferation and inhibiting apoptosis by targeting c-Myc
miR-181c	Tumor suppressor
miR-10b	Cell proliferation
miR-708	Induce carcinogenicity by down regulating Capase-2 level
miR-24	Enhance proliferation and reduces apoptosis

The deregulated salivary proteins and their association with OSCC are listed in Table 6.

Table 6. Deregulated salivary protein and their association with the OSCC

Protein Symbol	**Protein Name**	**Functional Importance in Cancer**
EDNI	Endothelin-1	Promote tumorigenesis
IL-8	Interleukin-8	Recruitment and activation of neutrophils, basophils, and T-cells which are in inflammatory response.
IL-1p	Interleukin 1 Beta	Inflammatory response, cell proliferation, differentiation, and cell survival
M2BP	Mac-2-binding protein	Modulation of cell-cell and cell-matrix interactions
CD59	CD59 Molecule	Activation of T cells.

MRP14	S100 Calcium Binding Protein A9	Recruitment, adhesion and migration of leukocytes. Induction of cytokine and chemokine secretion.
CCND1	Cyclin D1	Cell proliferation and metastasis
Ki67	Marker Of Proliferation Ki-67	Cell proliferation and metastasis
LDH	Lactate Dehydrogenase	Production of lactate from pyruvate under anaerobic conditions which is key feature of cancer cells.
MMP-9	Matrix Metallopeptidase 9	Migration, chronic inflammation, angiogenesis and metastasis
OGG1	8-Oxoguanine DNA Glycosylase	Repair of oxidative DNA damage
SERPINB5	Maspin or Serpin Family B Member 5	Inhibition of tumor angiogenesis
MMP 2	Matrix Metallopeptidase 2	Regulation of vascularization and metastasis
IL-6	Interleukin 6	Down regulation of apoptosis and induction of uncontrolled cell proliferation thorough inactivation of p53 tumor suppressor gene by hypermethylation of its promoter region
TNF-α	Tumor Necrosis Factor-Alpha	Stimulation of cell proliferation, inhibition of apoptosis and secretion of proinflammatory cytokines through transcriptional activation of NF-kB
IL-1α	Interleukin 1 Alpha	Inflammation and angiogenesis
VEGF-A	Vascular Endothelial Growth Factor A	Angiogenesis, endothelial cell proliferation and migration.
bFGF	Fibroblast Growth Factor 2 (Basic)	Endothelial cell proliferation, invasion, and angiogenesis progression

The different salivary omics biomarker involved in cancer diagnosis and treatment are listed in Table 7.

Table 7. Salivary omics biomarkers involved in cancer diagnosis and treatment is listed below

Omics Biomarkers in Saliva	Role in Oral Diagnosis and Treatment
Promoter Methylation in EDNRB	Progression of Oral cancer
Over expression of DUSP1, IL-1β, IL-8, OAZ1, and S100P genes at mRNA level	Early detection of oral cancer
Over expression of MMP-1 and MMP-9 at mRNA level	Progression of dysplasia to Oral cancer
Reduced level of miR200a	Indication of early metastasis
Loss of miR-31	Detection of recurrent OSCC
Elevation of miR-27b	Early detection of oral cancer
Overexpression of miR-21 and miR-184 and reduced expression of miR145	Early detection of oral cancer
Increased expression of Transferrin protein	Early detection of oral cancer
Overexpression of bFGF	Detection of recurrent OSCC
Up regulation of lactic acid, 5-hydroxymethyl uracil, phenylalanine, valine and phosphocholine	Diagnostic indicator of OSCC

The authors concluded that glycosylation is the important feature of tumor cells as they secrete glycoconjugates into the body fluids [87].

Yang et al (2018) conducted a cross sectional study on OSCC patients to study the microbial community associated with the carcinoma staging and pathways associated with it. *F. nucleatum* was significantly increased in stage 1(1.66%) to stage 2 and 3(2.41%) to about stage 4 (3.31%). In early stages there was no specific operational taxonomic units. In stage 4 *Tannerella forsythia*, which is a Gram-negative anaerobic bacterium and a periodontal pathogen was found to be the unique pathogen in OSCC patients [88].

Irimie AI et al (2015) unveiled the novel technologies in the field of genomics, transcriptomics or proteomics for diagnosis of OSCC. There was an epigenetic alteration of NOTCH4 signaling which may represent a probable methylation signature for OSCC recurrence. Methylation alterations was also found in HPV-driven patients, but was of less significance [(89)].

Jancsik VA et al (2014) conducted a study to compare the proteomic profile of diabetic and healthy volunteers mainly focusing on the salivary biomarkers of OSCC. They concluded that there was only increase in the level of Annexin A8. Peroxiredoxin-2 seem to be correlated with the premalignant state type-2 diabetes which should not be underestimated for their role in development of OSCC [(90)].

Torralba MG et al (2021) conducted a study on oral microbiome deep Illumina sequencing in conjunction with the microbial community associated with the proteomics. This study suggested that saliva may not only reflect the microbiota associated with the tumor tissue and can also be used as a target specimen for diagnosis of oral cancer [(91)].

Hu X et al (2021) explained the relationship between microflora associated with the periodontitis and oral cancer cells. The authors suggested that monitoring the changes in the oral microbiota may be helpful in early detection of oral cancer. The ecological bacteria may be used as auxiliary screening method for OSCC [(92)].

Chen J et al (2017) elucidated the relationship between microbial alterations in oral cancer, pancreatic cancer, lung cancer, and colorectal cancer. The authors stated that identification of single carcinogenic microbe such as *Enterotoxigenic Bacteroides fragilis (*ETBF) must be investigated. Studies must confirm that the person is exposed to a specific microbe of interest before the onset of disease and approaches has to be made for prevention of disease through vaccination of that implicated microbe [(93)].

IMPACT OF LIFESTYLE AND ENVIRONMENTAL FACTORS

Influence of smoking and alcohol on oral microbiota and oscc

Smoking and alcohol consumption are well-known risk factors for oral squamous cell carcinoma (OSCC), and their influence on the oral microbiota further elucidates their role in carcinogenesis. Both habits induce significant changes in the composition and function of oral microbial communities, contributing to an environment that can promote cancer development.

Galvin et al (2023) observed that the tongue and floor of mouth are found to be the high-risk sites for OSCC. The authors concluded that smoking was found to be the major risk factor for development of OSCC as it influences the oral microbiome by depleting common microbiome of the oral cavity [(94)].

Saxena et al (2022) conducted a study to reveal the microbial community associated with the healthy and oral cancer patients using smokeless tobacco. They concluded that there was a possible role of smokeless tobacco consumption for the transition of healthy oral microbiome to inflammation associated oral microbiome. *Streptococcus* genus was to found to be significantly abundant in healthy individuals compared to OSCC patients [(95)].

Diet, nutrition and their role in microbial dysbiosis

Diet and nutrition play crucial roles in shaping the oral microbiome, and imbalances can lead to microbial dysbiosis, which is associated with various health issues, including oral diseases and potentially oral squamous cell carcinoma (OSCC). Understanding how different dietary components influence the oral microbiota is key to developing nutritional strategies for maintaining oral health and preventing disease.

Santonocito et al (2022) explained the influence of diet on the oral microbiome and the creation of a suitable microenvironment for the development of periodontal disease. The authors concluded that diet plays an important role in oral dysbiosis of periodontal disease, as it provides nutritional supplement for microorganisms [(96)].

Paola Zanetta et al (2021) described the impact of vitamin D and oral microbiota on oral cancer. The authors stated that vitamin D can directly control the immune system via vitamin D receptors (VDR) in both activated or naïve CD4 and CD8 T cell, B cells which indirectly affects the gut microflora [(97)].

CHEMOTHERAPEUTIC AGENTS IN ORAL SQUAMOUS CELL CARCINOMA

Chemotherapeutic agents play a critical role, especially in advanced stages of the disease, in reducing tumor size, preventing metastasis, and improving survival rates. Here are some of the main chemotherapeutic agents used in OSCC management.

1. **Cisplatin**:

    - ✓ Cisplatin is an effective chemotherapy drug used for OSCC.
    - ✓ **Mechanism of Action:** Cisplatin forms cross-links with DNA, preventing replication and transcription, leading to apoptosis.
    - ✓ It is often administered intravenously (IV) and can be given alone or in combination with other treatments like radiation therapy
    - ✓ **Side Effects:** Nephrotoxicity, ototoxicity, neurotoxicity, and myelosuppression

2. **Carboplatin**:

    - ✓ Similar to cisplatin, carboplatin is another platinum-based chemotherapy drug.
    - ✓ **Mechanism of Action:** Similar to cisplatin, it forms DNA cross-links and induces apoptosis
    - ✓ It is used in combination with other therapies to treat OSCC.
    - ✓ **Side Effects:** Lower incidence of nephrotoxicity and ototoxicity compared to cisplatin, but still causes myelosuppression and gastrointestinal disturbances

3. **5-fluorouracil (5-FU)**:

    - ✓ 5-FU is a fluoropyrimidine drug that interferes with cancer cell growth.
    - ✓ **Mechanism of Action:** A pyrimidine analogue that inhibits thymidylate synthase, leading to disruption of DNA synthesis and function.
    - ✓ It can be administered orally or intravenously and is commonly used in OSCC treatment
    - ✓ **Side Effects:** Myelosuppression, mucositis, hand-foot syndrome, and gastrointestinal toxicity.

4. **Paclitaxel (Taxol)**:

    - ✓ Paclitaxel is a taxane chemotherapy drug.

- ✓ **Mechanism of Action:** Inhibits dihydrofolate reductase, leading to disruption of DNA synthesis and repair
- ✓ It is used to treat OSCC and can be given intravenously
- ✓ **Side Effects:** Neutropenia, neuropathy, hypersensitivity reactions, and myalgia

5. **Docetaxel (Taxotere)**:

- ✓ Docetaxel is another taxane drug used in OSCC management.
- ✓ **Mechanism of Action:** A taxane that promotes microtubule assembly while preventing their disassembly, leading to cell cycle arrest and apoptosis
- ✓ Like paclitaxel, it is administered intravenously.
- ✓ **Side Effects:** Neutropenia, neuropathy, fluid retention, and hypersensitivity reactions

6. **Hydroxyurea**:

- ✓ Hydroxyurea is an antimetabolite drug that interferes with cancer cell DNA synthesis.
- ✓ It may be used in combination with other therapies for OSCC treatment.

7. **Cetuximab**

- ✓ Mechanism of Action: A monoclonal antibody targeting the epidermal growth factor receptor (EGFR), inhibiting cell growth and survival signals.
- ✓ Often combined with radiation therapy or used with platinum-based chemotherapy for recurrent or metastatic OSCC.
- ✓ Side Effects: Skin rash, infusion reactions, hypomagnesemia, and gastrointestinal disturbances.

8. **Bleomycin**

- ✓ Mechanism of Action: Induces DNA strand breaks by generating free radicals, leading to apoptosis.
- ✓ Occasionally used in combination with other chemotherapeutic agents for OSCC.
- ✓ Side Effects: Pulmonary fibrosis, skin changes, and mucositis.

9. Capecitabine

- ✓ Mechanism of Action: An oral prodrug that is metabolized to 5-FU, exerting similar cytotoxic effects.
- ✓ Used in combination with other chemotherapeutic agents for advanced OSCC.
- ✓ Side Effects: Similar to 5-FU, including myelosuppression, mucositis, and gastrointestinal toxicity [(98)].

Nanoparticle drug delivery systems are emerging as a promising approach for the management of oral squamous cell carcinoma (OSCC). These systems leverage the unique properties of nanoparticles to enhance the efficacy, specificity, and safety of anticancer therapies.

Types of Nanoparticles Used in OSCC

1. **Magnetic Nanoparticles (MNPs)**: These NPs have been investigated for their potential in treating drug-resistant cancer cells. MNPs can be used for both drug delivery and imaging purposes.
2. **Liposomes**: Liposomes are lipid-based NPs that can encapsulate drugs and deliver them to specific sites. They offer controlled drug release and reduced side effects.
3. **Polymeric Nanoparticles**: These NPs are made from biocompatible polymers and can be tailored for targeted drug delivery.
4. **Gold (Au) Nanoparticles**: Au NPs have unique properties and can be functionalized for targeted therapy.
5. **Nano-Diamonds**: Nano-diamonds have emerged as a novel platform for drug delivery and imaging due to their biocompatibility and stability[(99)].

Manipulating the oral microbiome to restore a healthy balance or to eliminate pathogenic bacteria could have therapeutic benefits.

Mohd Fuah et al (2022) described the mechanism of probiotics. Prebiotics in management of oral cancer was found to be a novel approach for prevention of oral cancer.

Probiotics is defined as "living microorganisms which when administered in adequate amounts confer a health benefit on the host". This living microorganism can help maintain a healthy oral environment and potentially reduce carcinogenic bacterial load. The important organism used for this probiotic preparation include *Lactobacillus, Bifidobacterium, Saccharomyces, Streptococcus, Enterococcus, Escherichia,* and *Bacillus*.

Prebiotics is the substance present in food especially oligosaccharides which has the ability to stimulate the growth of beneficial bacteria or fungi. Most of the prebiotics is a subset of carbohydrates like galacto-oligosaccharides, fructo-oligosaccharides which trigger the increase of brain-derived neurotrophic factor (BDNF) and N-methyl-d-aspartate receptor (NMDAR) subunits, which help to repair failing brain cells and protect healthy brain cell.

Synbiotics is defined as the combination of both probiotics and prebiotics. By preserving the viability of healthy oral bacteria in the mouth cavity, the use of synbiotics may also help to maintain the oral microbiome by avoiding the colonization of oral pathogens such *Streptococcus mutans* and *Candida spp.*

Postbiotics is defined as the preparation of inanimate microorganisms, their components that confer a health benefit on host. The variety of substances and components used for the preparation of postbiotics include teichoic acid, muropeptides generated from peptidoglycan, functional proteins, secreted polysaccharides, extracellular polysaccharides (EPS), microbial fractions, and pili-type structure. Postbiotics are thought to provide therapeutic and preventive properties comparable to probiotics, but at a lower risk, especially for individuals with impaired immune systems.

This review also emphasizes the paucity of research on the use of postbiotics and synbiotics in the treatment of oral cancer. It is highly desired to conduct more research on their cytotoxicity, immunomodulatory, and cancer-preventive activities. Supplemental therapies such as probiotics, prebiotics, synbiotics, and postbiotics can help lessen the negative effects of cancer treatments. However, further research is needed to fully understand the substance's efficacy and effectiveness. The great potential of these novel cancer and infectious disease therapy techniques makes them worthwhile to investigate overall [(100)].

Cui S et al (2023) conducted a review on nanoparticle drug delivery system for OSCC management and summarized the recent advances in application of major nanoparticles as novel drug delivery system [(101)].

Surer et al (2021) conducted invitro cell line study with the gold nanoparticles which was synthesized and conjugated with cetuximab and cisplatin. They concluded that this nanodrug complex has higher potential to increase the cytotoxicity and also has increased resistance to radiotherapy in the treatment of oral cancer [(102)].

Abdel Hamid et al (2021) conducted research on 5- fluorouracil (5Fu), camptothecin (CPT), and a fibroblast growth factor receptor1-inhibitor (FGFR1i) to gold nanospheres (AuNSs) and followed their trajectories in Syrian hamsters with chemically induced buccal carcinomas. They concluded that the findings suggest that the biological processes occurring within the cell do not foretell the result observed in our in vivo model. Additionally, our findings imply that AuNSs, rather than powerful anticancer medications, preferentially increase the therapeutic impact of small molecule inhibitors like FGFR1i [(103)].

Bharadwaj R et al (2020) conducted a study on treatment of OSCC by using paclitaxel (PTX), 5-fluorouracil (5-FU) and ascorbic acid (AA) loaded solid lipid nanoparticles (SLN). The authors stimulated the oral cancer pathology in the animal model by using 4-NQO (4-nitroquinolone 1-oxide). They concluded that PTX, 5-FU and AA can be successfully entrapped within the SLN by using high-speed homogenization and ultrasonication method and can be used for OSCC treatment effectively [(104)].

Gusti-Ngurah-Putu et al (2019) reported the therapeutic efficacy of Encapsulating cisplatin (CDDP) into liposomes to form lipid-platinum-chloride nanoparticles (LPC NPs) in combination with the photodynamic therapy (PDT) on a xenograft model of OSCC. PDT+LPC prolongs the inhibition of tumour growth leading to minimal chemotherapy. The results indicate that a potent additive effect on the efficacy of chemotherapy is provided by PDT cytotoxicity. Therefore, the combination of PDT with LPC NPs have been shown to be more effective in treating the disease [(103)].

Shi et al (2018) conducted an in vivo study to assess the combined effect of TH287 and MDR1 siRNA therapeutic efficacy of OSCC in a xenograft model. The main reason for the failure of drug therapy is mainly due to multi-drug resistance (MDR) protein 1 (MDR1) or P-glycoprotein 1. TH287 is one among the protein MutT homolog 1(MTH1) that causes suppression of cancer cells. SiRNA is a 20–25 base pairs of RNA molecules that actively involves in the regulation of specific gene expression by controlling the gene silencing activity. They concluded that the combination of MDR1 siRNA with TH287 is effective and has increased anticancer efficacy [(106)].

Tanaka et al (2019) evaluated the anticancerous effect of novel platinum nanocomposite beads (PtNCP beads) and ovel platinum nanoparticles-conjugated latex beads (P2VPs) in Oral squamous cell carcinoma cell lines, HSC-3-M3 cells. The authors concluded that the PtNCP beads has increased anti-cancer effect and can be used as a drug transport carrier [(107)].

Kurakula et al (2020) conducted study to analyse the chitosan-coated Simvastatin (SIM) doped with quercetin (QRC) nanoparticles (NPs) loaded in an in-situ gel (ISG) formulation via a face-centred central composite design (FCCCD) to overcome the potential drawbacks of systemic chemotherapy for carcinoma of tongue. The controlled release of SIM from SIM–QRC NP-loaded ISG can be expected to induce or helps in the induction of angiogenesis process. ISG formulation showed a considerable increase in apoptosis occurrence through caspase-3-mediation and it also enhances the tumor suppressor protein levels. All these features make SIM–QRC NP-loaded ISG a novel formulation which can be used as a new way to treat carcinomas [(108)].

Qiu et al (2020) observed that mesenchymal stem cell-derived exosome (MSC-EXO) can be used as a carrier to deliver Cabazitaxel (CTX). It is a taxane drug which induces apoptosis or autophagy and are sensitive to some drug resistant and concluded MSCT-EXO-based CTX delivery system can be used as a method for anticancer treatment [(109)].

Antimicrobial Therapies: play a significant role in the management of oral squamous cell carcinoma (OSCC) by addressing the infections that can complicate cancer treatment and affect overall patient outcomes. These therapies can help manage infections related to cancer, its treatment, and surgery. Targeted antimicrobial treatments can selectively reduce harmful bacteria implicated in OSCC, thereby potentially slowing down or preventing cancer progression.

Liang et al (2021) conducted research on carrimycin, a synthesized macrolide antibiotic. In this study, the authors concluded that on increasing the concentration of carrimycin, the proliferation, colony formation and the ability of migration of OSCC cells was inhibited and tumor growth was arrested [(110)].

In a review by Liu et al (2019), epidermal growth factor receptor (EGFR) and human epidermal growth factor (HER/ErBb) receptor tyrosine kinase are overexpressed in more than 80% of the HNSCC. The two drugs that has the ability to recognize this extracellular ligand binding protein and interfere with the activation of the receptors are cetuximab and nimotuzumab. Cetuximab is a IgG1 monoclonal antibody derived from chimeric human mouse. Nimotuzumab is a humanized IgG1 monoclonal antibody drug with about 95% of humanization [(111)].

In a review by Radiac et al (2021) on orolome and antimicrobial therapeutics for the treatment of oral cancer, the antimicrobial peptide has the ability to specifically bind to tumors without causing damage to normal tissue. The epidrugs are used to inhibit or modulate the epigenetic proteins associated with the disease [112].

CONCLUSION

The oral microbiome plays a significant role in the development and progression of oral squamous cell carcinoma (OSCC). Studies have demonstrated that specific microbial communities and dysbiosis in the oral cavity are associated with OSCC. The presence of certain bacteria, such as *Fusobacterium nucleatum*, *Porphyromonas gingivalis,* and *Treponema denticola*, has been linked to increased inflammation and carcinogenesis, suggesting that these pathogens may contribute to the malignant transformation of oral epithelial cells.

Moreover, the interaction between oral pathogens and the host immune response appears to be crucial in the pathogenesis of OSCC. Bacteria can induce chronic inflammation, which creates a microenvironment conducive to cancer development. This inflammation is often characterized by the production of pro-inflammatory cytokines, reactive oxygen species, and other mediators that can damage cellular DNA, promote mutations, and facilitate tumor growth.

The analysis of the oral microbiome as a diagnostic and prognostic tool for OSCC is promising. Changes in microbial composition could potentially serve as biomarkers for early detection and risk assessment. Therapeutic strategies targeting the oral microbiome, such as probiotics, antibiotics, or antimicrobial peptides, are being explored to prevent or treat OSCC.

In conclusion, the oral microbiome is a critical factor in the etiology of OSCC. Understanding the intricate relationships between oral microbes and the host's cellular environment will enhance our ability to develop targeted interventions and improve clinical outcomes for patients with OSCC. Future research should focus on elucidating the mechanisms by which specific bacteria contribute to oral carcinogenesis and exploring microbiome-based therapeutic approaches.

REFERENCES

1. Brown JS, Amend SR, Austin RH, Gatenby RA, Hammarlund EU, Pienta KJ. Updating the Definition of Cancer. Mol Cancer Res. 2023 Nov 1;21(11):1142-1147. doi: 10.1158/1541-7786.MCR-23-0411. PMID: 37409952; PMCID: PMC10618731.
2. 2.Chow LQM. Head and Neck Cancer. Longo DL, editor. New England Journal of Medicine. 2020 Jan 2;382(1):60–72
3. Johnson DE, Burtness B, Leemans CR, Lui VWY, Bauman JE, Grandis JR. Head and neck squamous cell carcinoma. Nature reviews Disease primers. 2020 Nov 26;6(1):92
4. La Rosa GRM, Gattuso G, Pedullà E, Rapisarda E, Nicolosi D, Salmeri M. Association of oral dysbiosis with oral cancer development. Oncol Lett. 2020 Apr;19(4):3045-3058. doi: 10.3892/ol.2020.11441. Epub 2020 Mar 3. PMID: 32211076; PMCID: PMC7079586
5. Deo PN, Deshmukh R. Oral microbiome and oral cancer – The probable nexus. J Oral Maxillofac Pathol 2020;24:361-7
6. Sarode G, Sarode SC, Tupkari J, Patil S. Is oral squamous cell carcinoma unique in terms of intra and intertumoral heterogeneity? Transl Res Oral Oncol 2017;2:16
7. Sami A, Elimairi I, Stanton C, Ross RP, Ryan CA. The Role of the Microbiome in Oral Squamous Cell Carcinoma with Insight into the Microbiome–Treatment Axis. International Journal of Molecular Sciences. 2020 Oct 29;21(21):8061
8. Baker, J.L., Mark Welch, J.L., Kauffman, K.M. *et al.* The oral microbiome: diversity, biogeography and human health. *Nat Rev Microbiol* **22**, 89–104 (2024). https://doi.org/10.1038/s41579-023-00963-6
9. Torralba, M.G., Aleti, G., Li, W. *et al.* Oral Microbial Species and Virulence Factors Associated with Oral Squamous Cell Carcinoma. *Microb Ecol* **82**, 1030–1046 (2021). https://doi.org/10.1007/s00248-020-01596-5
10. Marchesi, J.R.; Ravel, J. The Vocabulary of Microbiome Research: A Proposal. Microbiome 2015, 3, 31.
11. Cho, I.; Blaser, M.J. The Human Microbiome: At the Interface of Health and Disease. Nat. Rev. Genet. 2012, 13, 260–270.
12. Pignatelli, P.; Romei, F.M.; Bondi, D.; Giuliani, M.; Piattelli, A.; Curia, M.C. Microbiota and Oral Cancer as A Complex and Dynamic Microenvironment: A Narrative Review from Etiology to Prognosis. Int. J. Mol. Sci. 2022, 23, 8323

13. Schmidt, B.L.; Kuczynski, J.; Bhattacharya, A.; Huey, B.; Corby, P.M.; Queiroz, E.L.S.; Nightingale, K.; Kerr, A.R.; DeLacure, M.D.; Veeramachaneni, R.; et al. Changes in Abundance of Oral Microbiota Associated with Oral Cancer. PLoS ONE 2014, 9, e98741.
14. Lamont, R.J.; Koo, H.; Hajishengallis, G. The Oral Microbiota: Dynamic Communities and Host Interactions. Nat. Rev. Microbiol. 2018, 16, 745–759.
15. Ly, M.; Abeles, S.R.; Boehm, T.K.; Robles-Sikisaka, R.; Naidu, M.; Santiago-Rodriguez, T.; Pride, D.T. Altered Oral Viral Ecology in Association with Periodontal Disease. mBio 2014, 5, e01133-14.
16. Khatoon, J. Role of Helicobacter Pylori in Gastric Cancer: Updates. World J. Gastrointest. Oncol. 2016, 8, 147
17. Koshiol, J.; Wozniak, A.; Cook, P.; Adaniel, C.; Acevedo, J.; Azócar, L.; Hsing, A.W.; Roa, J.C.; Pasetti, M.F.; Miquel, J.F.; et al. Salmonella Enterica Serovar Typhi and Gallbladder Cancer: A Case–Control Study and Meta-analysis. Cancer Med. 2016, 5, 3310–3325.
18. Nosho, K. Association of Fusobacterium Nucleatum with Immunity and Molecular Alterations in Colorectal Cancer. World J. Gastroenterol. 2016, 22, 557.
19. Pignatelli, P.; Iezzi, L.; Pennese, M.; Raimondi, P.; Cichella, A.; Bondi, D.; Grande, R.; Cotellese, R.; Di Bartolomeo, N.; Innocenti, P.; et al. The Potential of Colonic Tumor Tissue Fusobacterium Nucleatum to Predict Staging and Its Interplay with Oral Abundance in Colon Cancer Patients. Cancers 2021, 13, 1032.
20. Karpi ´nski, T. Role of Oral Microbiota in Cancer Development. Microorganisms 2019, 7, 20. [CrossRef] 18. Willis, J.R.; Gabaldón, T. The Human Oral Microbiome in Health and Disease: From Sequences to Ecosystems. Microorganisms 2020, 8, 308.
21. Chattopadhyay, I.; Verma, M.; Panda, M. Role of Oral Microbiome Signatures in Diagnosis and Prognosis of Oral Cancer. Technol. Cancer Res. Treat. 2019, 18, 153303381986735.
22. Aas, J.A.; Paster, B.J.; Stokes, L.N.; Olsen, I.; Dewhirst, F.E. Defining the Normal Bacterial Flora of the Oral Cavity. J. Clin. Microbiol. 2005, 43, 5721–5732
23. Pathak, J.L.; Yan, Y.; Zhang, Q.; Wang, L.; Ge, L. The Role of Oral Microbiome in Respiratory Health and Diseases. Respir. Med. **2021**, 185, 106475.
24. Lo Giudice, R.; Militi, A.; Nicita, F.; Bruno, G.; Tamà, C.; Lo Giudice, F.; Puleio, F.; Calapai, F.; Mannucci, C. Correlation between Oral Hygiene and IL-6 in Children. Dent. J. **2020**, 8, 91
25. Kozak, M.; Pawlik, A. The Role of the Oral Microbiome in the Development of Diseases. Int. J. Mol. Sci. **2023**, 24, 5231.

26. Gomez, A.; Nelson, K.E. The Oral Microbiome of Children: Development, Disease, and Implications Beyond Oral Health. Microb. Ecol. **2017**, 73, 492–503.
27. D'Ambrosio, F.; Santella, B.; Di Palo, M.P.; Giordano, F.; Lo Giudice, R. Characterization of the Oral Microbiome in Wearers of Fixed and Removable Implant or Non-Implant-Supported Prostheses in Healthy and Pathological Oral Conditions: A Narrative Review. Microorganisms **2023**, 11, 1041.
28. Baraniya D, Chen T, Nahar A, et al. Supragingival mycobiome and inter-kingdom interactions in dental caries. J Oral Microbiol.2020;12(1):1729305.
29. Diaz PI, Dongari- Bagtzoglou A. Critically appraising the significance of the oral mycobiome. J Dent Res. 2021;100(2):133-140.
30. He, J.; Li, Y.; Cao, Y.; Xue, J.; Zhou, X. The Oral Microbiome Diversity and Its Relation to Human Diseases. Folia Microbiol. 2015, 60, 69–80.
31. Prasad, R.B.K.; Sharma, A.; Babu, H.M. An Insight into Salivary Markers in Oral Cancer. Dent. Res. J. 2013, 10, 9.
32. Kilian M, Chapple IL, Hannig M, et al. The oral microbiome-an update for oral healthcare professionals. Br Dent J. 2016;221(10):657-666. doi: 10.1038/sj.bdj.2016.865
33. Gao L, Xu T, Huang G, Jiang S, Gu Y, Chen F. Oral microbiomes: more and more importance in oral cavity and whole body. Protein Cell. 2018;9(5):488-500. doi: 10.1007/s13238-018-05481
34. Deo PN, Deshmukh R. Oral microbiome: Unveiling the fundamentals. JOMFP. 2019;23(1):122-128. doi: 10.4103/jomfp.JOMFP_304_18
35. Zarco MF, Vess TJ, Ginsburg GS. The oral microbiome in health and disease and the potential impact on personalized dental medicine. Oral Dis. 2012;18(2):109- 120. doi: 10.1111/j.1601-0825.2011.01851.x
36. Demmitt BA, Corley RP, Huibregtse BM, et al. Genetic influences on the human oral microbiome. BMC Genom. 2017;18(1):659. doi: 10.1186/s12864-017- 4008-8
37. O'Grady I, Anderson A, O'Sullivan J. The interplay of the oral microbiome and alcohol consumption in oral squamous cell carcinomas. Oral Oncol. 2020;110:105011. doi: 10.1016/j. oraloncology.2020.105011
38. Verma D, Garg PK, Dubey AK. Insights into the human oral microbiome. Arch Microbiol. 2018;200(4):525- 540. doi: 10.1007/s00203-018-1505-3
39. Sultan AS, Kong EF, Rizk AM, Jabra-Rizk MA. The oral microbiome: A Lesson in coexistence. PLoS Pathog. 2018;14(1):e1006719. doi: 10.1371/journal. ppat.1006719

40. Rasmussen LT, Labio RW, Gatti LL, et al. Helicobacter pylori detection in gastric biopsies, saliva and dental plaque of Brazilian dyspeptic patients. Mem Inst Oswaldo Cruz. 2010;105(3):326-330. doi: 10.1590/ S0074-02762010000300015
41. Laheij AM, de Soet JJ, von dem Borne PA, et al. Oral bacteria and yeasts in relationship to oral ulcerations in hematopoietic stem cell transplant recipients. Support Care Cancer. 2012;20(12):3231-3240. doi: 10.1007/ s00520-012-1463-2
42. Paster BJ, Boches SK, Galvin JL, et al. Bacterial diversity in human subgingival plaque. J Bacteriol. 2001;183(12):3770-3783. doi: 10.1128/ JB.183.12.3770-3783.2001
43. Awano S, Ansai T, Takata Y, et al. Oral health and mortality risk from pneumonia in the elderly. J Dent Res. 2008;87(4):334-339. doi: 10.1177/154405910808700418
44. Dewhirst FE, Chen T, Izard J, et al. The human oral microbiome. J Bacteriol. 2010;192(19):5002-5017. doi: 10.1128/JB.00542-10
45. Dzidic M, Collado MC, Abrahamsson T, et al. Oral microbiome development during childhood: an ecological succession influenced by postnatal factors and associated with tooth decay. ISME J. 2018;12(9):2292-2306. doi: 10.1038/s41396-018- 0204-z
46. Zaura E, Nicu EA, Krom BP, Keijser BJ. Acquiring and maintaining a normal oral microbiome: current perspective. Front Cell Infect Microbiol. 2014 Jun 26;4:85. doi: 10.3389/fcimb.2014.00085. PMID: 25019064; PMCID: PMC4071637.
47. Sedghi L, DiMassa V, Harrington A, Lynch SV, Kapila YL. The oral microbiome: Role of key organisms and complex networks in oral health and disease. Periodontology 2000. 2021;87:107131.
48. Pignatelli, P.; Romei, F.M.; Bondi, D.; Giuliani, M.; Piattelli, A.; Curia, M.C. Microbiota and Oral Cancer as A Complex and Dynamic Microenvironment: A Narrative Review from Etiology to Prognosis. Int. J. Mol. Sci. 2022, 23, 8323. https://doi.org/10.3390/ ijms23158323
49. Theis, K.R.; Dheilly, N.M.; Klassen, J.L.; Brucker, R.M.; Baines, J.F.; Bosch, T.C.G.; Cryan, J.F.; Gilbert, S.F.; Goodnight, C.J.; Lloyd, E.A.; et al. Getting the Hologenome Concept Right: An Eco-Evolutionary Framework for Hosts and Their Microbiomes. mSystems 2016, 1, e00028-16
50. Shokeen, B.; Dinis, M.D.B.; Haghighi, F.; Tran, N.C.; Lux, R. Omics and Interspecies Interaction. Periodontol. 2000 2020, 85, 101–111
51. Sung, H.; Ferlay, J.; Siegel, R.L.; Laversanne, M.; Soerjomataram, I.; Jemal, A.; Bray, F. Global Cancer Statistics 2020: GLOBOCAN Estimates of Incidence and Mortality Worldwide for 36 Cancers in 185 Countries. CA Cancer J. Clin. **2021**, 71, 209–249.

52. Ferlay, J.; Ervik, M.; Lam, F.; Colombet, M.; Mery, L.; Piñeros, M.; Znaor, A.; Soerjomataram, I.; Bray, F. Global Cancer Observatory: Cancer Today; International Agency for Research on Cancer: Lyon, France, 2020.
53. Cheong, S.C.; Vatanasapt, P.; Yi-Hsin, Y.; Zain, R.B.; Kerr, A.R.; Johnson, N.W. Oral cancer in South East Asia. Transl. Res. Oral Oncol. **2017**, 2, 2057178X1770292
54. Gormley, M.; Creaney, G.; Schache, A.; Ingarfield, K.; Conway, D.I. Reviewing the epidemiology of head and neck cancer: Definitions, trends and risk factors. Br. Dent. J. **2022**, 233, 780–786.
55. Fitzmaurice, C.; Allen, C.; Barber, R.M.; Barregard, L.; Bhutta, Z.A.; Brenner, H.; Dicker, D.J.; Chimed-Orchir, O.; Dandona, R.; Dandona, L.; et al. Global, Regional, and National Cancer Incidence, Mortality, Years of Life Lost, Years Lived with Disability, and Disability-Adjusted Life-Years for 32 Cancer Groups, 1990 to 2015: A Systematic Analysis for the Global Burden of Disease Study Global Burden of Disease Cancer Collaboration. JAMA Oncol. **2017**, 3, 524–548.
56. Bosetti, C.; Carioli, G.; Santucci, C.; Bertuccio, P.; Gallus, S.; Garavello, W.; Negri, E.; La Vecchia, C. Global trends in oral and pharyngeal cancer incidence and mortality. Int. J. Cancer **2020**, 147, 1040–1049.
57. Miranda-Filho, A.; Bray, F. Global patterns and trends in cancers of the lip, tongue and mouth. Oral Oncol. **2020**, 102, 104551.
58. Kawakita, D.; Oze, I.; Iwasaki, S.; Matsuda, T.; Matsuo, K.; Ito, H. Trends in the incidence of head and neck cancer by subsite between 1993 and 2015 in Japan. Cancer Med. **2022**, 11, 1553–1560.
59. Louie, K.S.; Mehanna, H.; Sasieni, P. Trends in head and neck cancers in England from 1995 to 2011 and projections up to 2025. Oral Oncol. **2015**, 51, 341–348.
60. Creaney, G.; McMahon, A.D.; Ross, A.J.; Bhatti, L.A.; Paterson, C.; Conway, D.I. Head and neck cancer in the UK: What was the stage before COVID-19? UK cancer registries analysis (2011–2018). Br. Dent. J. **2022**, 233, 787–793.
61. Bagal S, Budukh A, Thakur JS, Dora T, Qayyumi B, Khanna D, Fernandes D, Chakravarti P, Singh R, Patil S, Dikshit R, Chaturvedi P. Head and neck cancer burden in India: an analysis from published data of 37 population-based cancer registries. Ecancermedicalscience. 2023 Sep 21;17:1603. doi: 10.3332/ecancer.2023.1603. PMID: 37799939; PMCID: PMC10550331.
62. Tandon P, Dadhich A, Saluja H, Bawane S, Sachdeva S. The prevalence of squamous cell carcinoma in different sites of oral cavity at our Rural Health Care Centre in Loni,

Maharashtra - a retrospective 10-year study. Contemp Oncol (Pozn). 2017;21(2):178-183. doi: 10.5114/wo.2017.68628. Epub 2017 Jun 30. PMID: 28947890; PMCID: PMC5611509
63. Mummudi N, Agarwal JP, Chatterjee S, Mallick I, Ghosh-Laskar S. Oral Cavity Cancer in the Indian Subcontinent – Challenges and Opportunities. Clinical Oncology. 2019 Aug;31(8):520–8.
64. Thankappan, Simi; Nedumpillil, Sherin1; Thomas, Valsa. Clinicopathologic features and risk factors associated with oral squamous cell carcinoma (OSCC): A comprehensive descriptive study analyzing the burden of OSCC in a tertiary-level hospital in North Kerala, India. Indian Journal of Cancer 60(4):p 534-541, Oct–Dec 2023. | DOI: 10.4103/ijc.IJC_203_21
65. Thomas, C.; Minty, M.; Vinel, A.; Canceill, T.; Loubières, P.; Burcelin, R.; Kaddech, M.; Blasco-Baque, V.; Laurencin-Dalicieux, S. Oral Microbiota: A Major Player in the Diagnosis of Systemic Diseases. Diagnostics 2021, 11, 1376. https:// doi.org/10.3390/diagnostics11081376
66. Gao L, Xu T, Huang G, Jiang S, Gu Y, Chen F. Oral microbiomes: more and more importance in oral cavity and whole body. Protein Cell. 2018 May;9(5):488-500. doi: 10.1007/s13238-018-0548-1. Epub 2018 May 7. PMID: 29736705; PMCID: PMC5960472.
67. Gazdeck, R.K.; Fruscione, S.R.; Adami, G.R.; Zhou, Y.; Cooper, L.F.; Schwartz, J.L. Diversity of the Oral Microbiome between Dentate and Edentulous Individuals. Oral Dis. 2019, 25, 911–918
68. Hubbers CU, Akgul B. HPV and cancer of the oral cavity.Virulence.2015;6(3):244-248.
69. Adler I, Muino A, Aguas S, et al. Helicobacter pylori and oral pathology: relationship with the gastric infection. World J Gastroenterol.2014;20(29):9922-9935.
70. Meng X, Wang Q, He C, et al. An inverse association of Helicobacter pylori infection with oral squamous cell carcinoma. J Oral Pathol Med. 2016;45(1):17-22
71. Belibasakis GN, Senevirantne CJ,Jayasinghe RD, Vo P-D, Bostanci N, Choi Y. Bacteriome and mycobiome dysbiosis in oral mucosal dysplasia and oral cancer. Periodontol 2000. 2024;00:1-17. doi:10.1111/prd.12558
72. Lamont RJ, Fitzsimonds ZR, Wang H, Gao S. Role of Porphyromonas gingivalis in oral and orodigestive squamous cell carcinoma.Periodontol 2000. 2022;89(1):154-165
73. Vyhnalova, T.; Danek, Z.;Gachova, D.; Linhartova, P.B. The Role of the Oral Microbiota in the Etiopathogenesis of Oral Squamous Cell Carcinoma. Microorganisms 2021,9, 1549. https://doi.org/10.3390/microorganisms9081549

74. Yu X, Shi Y, Yuan R, Chen Z, Dong Q, Han L, Wang L, Zhou J. Microbial dysbiosis in oral squamous cell carcinoma: A systematic review and meta-analysis. Heliyon. 2023 Jan 24;9(2):e13198. doi: 10.1016/j.heliyon.2023.e13198. PMID: 36793959; PMCID: PMC9922960.
75. Wang, J.; Gao, B. Mechanisms and Potential Clinical Implications of Oral Microbiome in Oral Squamous Cell Carcinoma. Curr. Oncol. 2024, 31, 168–182. https://doi.org/10.3390/curroncol31010011
76. Geng, F.; Liu, J.; Guo, Y.; Li, C.; Wang, H.; Wang, H.; Zhao, H.; Pan, Y. persistent exposure to Porphyromonas gingivalis promotes proliferative and invasion capabilities, and tumorigenic properties of human immortalized oral epithelial cells. Front. Cell Infect.Microbiol. 2017
77. Sayehmiri, F.; Sayehmiri, K.; Asadollahi, K.; Soroush, S.; Bogdanovic, L.; Jalilian, F.A.; Emaneini, M.; Taherikalani, M. The prevalence rate of Porphyromonas gingivalis and its association with cancer: A systematic review and meta-analysis. Int. J.Immunopathol. Pharmacol. 2015, 28, 160–167.
78. O'Shea, J.J.; Schwartz, D.M.; Villarino, A.V.; Gadina, M.; McInnes, I.B.; Laurence, A. The JAK-STAT pathway: Impact on human disease and therapeutic intervention. Annu. Rev. Med. 2015, 66, 311–32
79. Sukmana BI, Saleh RO, Najim MA, ALGhamdi HS, Achmad H, Al-Hamdani MM, Taher AAY, Alsalamy A, Khaledi M and Javadi K (2024) Oral microbiota and oral squamous cell carcinoma: a review of their relation and carcinogenic mechanisms. Front. Oncol. 14:1319777. doi: 10.3389/fonc.2024.1319777
80. Chattopadhyay I, Verma M, Panda M. Role of Oral Microbiome Signatures in Diagnosis and Prognosis of Oral Cancer. Technology in Cancer Research & Treatment. 2019 Jan;18:153303381986735
81. Kakabadze MZ, Paresishvili T, Karalashvili L, Chakhunashvili D, Kakabadze Z. Oral microbiota and oral cancer: Review. Oncol Rev. 2020 Jul 6;14(2):476. doi: 10.4081/oncol.2020.476. PMID: 32676172; PMCID: PMC7358985.
82. Monteiro JS, Kaushik K, de Arruda JAA,Georgakopoulou E, Vieira AT, Silva TA,Devadiga D, Anyanechi CE and Shetty S (2024)Fungal footprints in oral cancer: unveiling the oral mycobiome.Front. Oral. Health 5:1360340.doi: 10.3389/froh.2024.1360340
83. Verma AH, Richardson JP, Zhou C, Coleman BM, Moyes DL, Ho J, et al. Oral epithelial cells orchestrate innate type 17 responses to Candida albicans through the virulence factor candidalysin. Sci Immunol. (2017) 2(17):eaam8834. doi: 10.1126/sciimmunol.aam8834

84. Engku Nasrullah Satiman EAF, Ahmad H, Ramzi AB, Abdul Wahab R, Kaderi MA, Wan Harun WHA, et al. The role of Candida albicans candidalysin ECE1gene in oral carcinogenesis. J Oral Pathol Med. (2020) 49(9):835–41. doi: 10.1111/jop.13014
85. Di Cosola, M.; Cazzolla,A.P.; Charitos, I.A.; Ballini, A.Inchingolo, F.; Santacroce, L. Candida albicans and Oral Carcinogenesis. A Brief Review. J. Fungi 2021, 7, 476.https://doi.org/10.3390/jof7060476
86. Heng W, Wang W, Dai T, Jiang P, Lu Y, Li R, Zhang M, Xie R, Zhou Y, Zhao M, Duan N, Ye Z, Yan F, Wang X.2022.Oral Bacteriome and Mycobiome across Stages of Oral Carcinogenesis. Microbiol Spectr10:e02737-22.https://doi.org/10.1128/spectrum.02737-22
87. Chattopadhyay I, Panda M. Recent trends of saliva omics biomarkers for the diagnosis and treatment of oral cancer. Journal of Oral Biosciences. 2019 Jun;61(2):84–94.
88. Yang C-Y, Yeh Y-M, Yu H-Y, Chin C-Y, Hsu C-W, Liu H, Huang P-J, Hu S-N, Liao C-T, Chang K-P and Chang Y-L (2018) Oral Microbiota Community Dynamics Associated With Oral Squamous Cell Carcinoma Staging. Front. Microbiol. 9:862. doi: 10.3389/fmicb.2018.00862
89. Irimie AI, Braicu C, Cojocneanu-Petric R, Berindan-Neagoe I, Campian RS. Novel technologies for oral squamous carcinoma biomarkers in diagnostics and prognostics. Acta Odontol Scand. 2015 Apr;73(3):161-8. doi: 10.3109/00016357.2014.986754. Epub 2015 Jan 19. PMID: 25598447.
90. Jancsik VA, Gelencser G, Maasz G, Schmidt J, Molnar GA, Wittmann I, Olasz L, Mark L. Salivary proteomic analysis of diabetic patients for possible oral squamous cell carcinoma biomarkers. Pathol Oncol Res. 2014 Jul;20(3):591-5. doi: 10.1007/s12253-013-9736-8. Epub 2013 Dec 17. PMID: 24343817
91. Torralba MG, Aleti G, Li W, Moncera KJ, Lin YH, Yu Y, Masternak MM, Golusinski W, Golusinski P, Lamperska K, Edlund A, Freire M, Nelson KE. Oral Microbial Species and Virulence Factors Associated with Oral Squamous Cell Carcinoma. Microb Ecol. 2021 Nov;82(4):1030-1046. doi: 10.1007/s00248-020-01596-5. Epub 2020 Nov 6. Erratum in: Microb Ecol. 2020 Nov 20;: PMID: 33155101; PMCID: PMC8551143.
92. Hu X, Shen X, Tian J. The effects of periodontitis associated microbiota on the development of oral squamous cell carcinoma. Biochemical and Biophysical Research Communications. 2021 Oct;576:80–5.
93. Chen J, Domingue JC, Sears CL. Microbiota dysbiosis in select human cancers: Evidence of association and causality. Seminars in Immunology. 2017 Aug;32:25–34

94. Galvin S, Moran GP, Healy CM. Influence of site and smoking on malignant transformation in the oral cavity: Is the microbiome the missing link? Front Oral Health. 2023 Mar 23;4:1166037. doi: 10.3389/froh.2023.1166037. PMID: 37035251; PMCID: PMC10076759.
95. Saxena R, Prasoodanan P K V, Gupta SV, Gupta S, Waiker P, Samaiya A, Sharma AK, Sharma VK. Assessing the Effect of Smokeless Tobacco Consumption on Oral Microbiome in Healthy and Oral Cancer Patients. Front Cell Infect Microbiol. 2022 Mar 31;12:841465. doi: 10.3389/fcimb.2022.841465. PMID: 35433507; PMCID: PMC9009303.
96. Santonocito S, Giudice A, Polizzi A, Troiano G, Merlo EM, Sclafani R, Grosso G, Isola G. A Cross-Talk between Diet and the Oral Microbiome: Balance of Nutrition on Inflammation and Immune System's Response during Periodontitis. Nutrients. 2022 Jun 11;14(12):2426. doi: 10.3390/nu14122426. PMID: 35745156; PMCID: PMC9227938.
97. Paola Zanetta, Diletta Francesca Squarzanti, Rita Sorrentino, Roberta Rolla, Paolo Aluffi Valletti, Massimiliano Garzaro, Valeria Dell'Era, Angela Amoruso & Barbara Azzimonti (2021) Oral microbiota and vitamin D impact on oropharyngeal squamous cell carcinogenesis: a narrative literature review, Critical Reviews in Microbiology, 47:2, 224-239, DOI:10.1080/1040841X.2021.1872487
98. Treating Oral Cavity and Oropharyngeal Cancer [Internet]. Available from: https://www.cancer.org/content/dam/CRC/PDF/Public/8766.00.pdf
99. Bhattacharya S. Way to Cure Oral Squamous Cell Carcinoma with Theranostics and Nanoparticular Approaches [Internet]. Oral Cancer - Current Concepts and Future Perspectives. IntechOpen; 2022. Available from:http://dx.doi.org/10.5772/intechopen.100181
100. Mohd Fuad AS, Amran NA, Nasruddin NS, Burhanudin NA, Dashper S, Arzmi MH. The Mechanisms of Probiotics, Prebiotics, Synbiotics, and Postbiotics in Oral Cancer Management. Probiotics and Antimicrobial Proteins. 2022 Sep 1
101. Cui S, Liu H and Cui G (2023),Nanoparticles as drug delivery systems in the treatment of oral squamous cell carcinoma: current status and recent progression.Front. Pharmacol. 14:1176422.doi: 10.3389/fphar.2023.1176422
102. Sürer Şİ, Elçitepe TB, Akçay D, Daşkın E, Çalıbaşı Kocal G, Arıcan Alıcıkuş Z, Eskiizmir G, Yapıcı K, Başbınar Y. A Promising, Novel Radiosensitizer Nanodrug Complex for Oral Cavity Cancer: Cetuximab and Cisplatin-Conjugated Gold

Nanoparticles. Balkan Med J. 2021 Sep;38(5):278-286. doi: 10.5152/balkanmedj.2021.21013. PMID: 34462254; PMCID: PMC8880916.

103. Abdel Hamid, H.M., Darwish, Z.E., Elsheikh, S.M. *et al.* Following cytotoxic nanoconjugates from injection to halting the cell cycle machinery and its therapeutic implications in oral cancer. *BMC Cancer* **21**, 170 (2021). https://doi.org/10.1186/s12885-021-07849-x

104. Bharadwaj R, Sahu BP, Haloi J, Laloo D, Barooah P, Keppen C, Deka M, Medhi S. Combinatorial therapeutic approach for treatment of oral squamous cell carcinoma. Artif Cells Nanomed Biotechnol. 2019 Dec;47(1):572-585. doi: 10.1080/21691401.2019.1573176. Erratum in: Artif Cells Nanomed Biotechnol. 2020 Dec;48(1):1302. PMID: 30831033.

105. Gusti-Ngurah-Putu EP, Huang L, Hsu YC. Effective Combined Photodynamic Therapy with Lipid Platinum Chloride Nanoparticles Therapies of Oral Squamous Carcinoma Tumor Inhibition. J Clin Med. 2019 Dec 2;8(12):2112. doi: 10.3390/jcm8122112. PMID: 31810241; PMCID: PMC6947167.

106. Shi X-Lei, Li Y, Zhao L-Ming, Su L-Wang, Ding G, Delivery of MTH1 inhibitor (TH287) and MDR1 siRNA via hyaluronic acid-based mesoporous silica nanoparticles for oral cancers treatment, Colloids and Surfaces B: Biointerfaces(2018), https://doi.org/10.1016/j.colsurfb.2018.09.076

107. Tanaka M, Okinaga T, Iwanaga K, Matsuo K, Toyono T, Sasaguri M, Ariyoshi W, Tominaga K, Enomoto Y, Matsumura Y, Nishihara T. Anticancer effect of novel platinum nanocomposite beads on oral squamous cell carcinoma cells. J Biomed Mater Res B Appl Biomater. 2019 Oct;107(7):2281-2287. doi: 10.1002/jbm.b.34320. Epub 2019 Jan 28. PMID: 30689290.

108. Kurakula M, Naveen NR. In Situ Gel Loaded with Chitosan-Coated Simvastatin Nanoparticles: Promising Delivery for Effective Anti-Proliferative Activity against Tongue Carcinoma. Mar Drugs. 2020 Apr 9;18(4):201. doi: 10.3390/md18040201. PMID: 32283782; PMCID: PMC7231276.

109. Qiu Y, Sun J, Qiu J, Chen G, Wang X, Mu Y, Li K, Wang W. Antitumor Activity of Cabazitaxel and MSC-TRAIL Derived Extracellular Vesicles in Drug-Resistant Oral Squamous Cell Carcinoma. Cancer Manag Res. 2020 Oct 29;12:10809-10820. doi: 10.2147/CMAR.S277324. PMID: 33149686; PMCID: PMC7605918.

110. Liang S, Zhao T, Zhou Z, Ju W, Liu Y, Tan Y, et al. Anti-tumor effect of carrimycin on oral squamous cell carcinoma cells in vitro and in vivo. Translational oncology. 2021 Jun 1;14(6):101074–4.

111. Liu L, Chen J, Cai X, Yao Z, Huang J. Progress in targeted therapeutic drugs for oral squamous cell carcinoma. Surg Oncol. 2019 Dec;31:90-97. doi: 10.1016/j.suronc.2019.09.001. Epub 2019 Sep 5. PMID: 31550560.

112. Radaic A, Ganther S, Kamarajan P, Grandis J, Yom SS, Kapila YL. Paradigm shift in the pathogenesis and treatment of oral cancer and other cancers focused on the oralome and antimicrobial-based therapeutics. Periodontol 2000. 2021;87:76–93. https://doi.org/10.1111/prd.12388

FSC
www.fsc.org
MIX
Papier | Fördert gute Waldnutzung
FSC® C083411

Kolibri 360 GmbH
Ferdinand-Jühlke-Straße 7
99095 Erfurt, Deutschland
produktsicherheit@kolibri360.de

Druck:
CPI Druckdienstleistungen GmbH
im Auftrag der
Kolibri 360 GmbH
Ferdinand-Jühlke-Str. 7
99095 Erfurt